Ergebnisse der Mathematik und ihrer Grenzgebiete 31

A Series of Modern Surveys in Mathematics

Jacob Wolfowitz

Coding Theorems of Information Theory

Third Edition

Springer-Verlag
Berlin Heidelberg New York 1978

Jacob Wolfowitz
Department of Mathematics, University of Illinois
Urbana, IL 61801/USA

AMS Subject Classification (1970): 94-02, 94A15, 62B10

ISBN-13: 978-3-642-66824-1 e-ISBN-13: 978-3-642-66822-7
DOI: 10.1007/978-3-642-66822-7

Library of Congress Cataloging in Publication Data. Wolfowitz, Jacob, 1910—.
Coding theorems of information theory. (Ergebnisse der Mathematik und ihrer Grenzgebiete; 31). Bibliography: p. Includes index. 1. Statistical communication theory. 2. Coding theory. I. Title. II. Series. TK510.2.W64. 1978. 001.5'1. 77-28231

Softcover reprint of the hardcover 1rd edition 1978

2141/3140-543210

Dedicated
to the memory of
Abraham Wald

Preface

The objective of the present edition of this monograph is the same as that of earlier editions, namely, to provide readers with some mathematical maturity a rigorous and modern introduction to the ideas and principal theorems of probabilistic information theory. It is not necessary that readers have any prior knowledge whatever of information theory. The rapid development of the subject has had the consequence that any one book can now cover only a fraction of the literature. The latter is often written by engineers for engineers, and the mathematical reader may have some difficulty with it. The mathematician who understands the content and methods of this monograph should be able to read the literature and start on research of his own in a subject of mathematical beauty and interest.

The present edition differs from the second in the following: Chapter 6 has been completely replaced by one on arbitrarily varying channels. Chapter 7 has been greatly enlarged. Chapter 8 on semi-continuous channels has been drastically shortened, and Chapter 11 on sequential decoding completely removed. The new Chapters 11—15 consist entirely of material which has been developed only in the last few years. The topics discussed are rate distortion, source coding, multiple access channels, and degraded broadcast channels. Even the specialist will find a new approach in the treatment of these subjects. Many of the proofs are new, more perspicuous, and considerably shorter than the original ones.

The reader who wishes to reach the material in the last five chapters as soon as possible can do so by reading only Chapters 2, the first half of 3, and 7.

Professor James H. Stapleton kindly read Chapter 6 and helped remove some minor errors. I can only hope that the new material in the last five chapters contains no major errors. Certainly it must contain minor errors for which I beg the reader's indulgence.

I am grateful to the Institute of Advanced Study of the University of Illinois and the Office of Scientific Research of the U.S. Air Office for financial support during a part of the writing of the monograph.

University of Illinois, June, 1977 J. Wolfowitz

Contents

1. Heuristic Introduction to the Discrete Memoryless Channel

The spirit of the problems discussed in the present monograph can already be gleaned from a consideration of the discrete memoryless channel, to a heuristic discussion of which the present chapter is devoted. In this discussion there will occur terms not yet precisely defined, to which the reader should give their colloquial meaning. This procedure is compatible with the purpose of the present chapter, which is to motivate the problems to be discussed later in the book, and not to carry forward the theory logically. The reader scornful of such unmathematical behavior or in no need of motivation may proceed at once to Chapter 2 without any loss of logical continuity. Such definitions as are given in the present chapter will be repeated later.

We suppose that a stream of symbols, each of which is a letter of an alphabet, is being sent over a discrete, noisy, memoryless channel. "Discrete" (a term of engineering origin) means here that the alphabet of the letters being sent, and the alphabet of the letters being received (which need not be the same) each have a finite number (>1) of symbols. We take these two numbers to be the same, say a. Since the actual symbols will not matter we may assume, and do, that both alphabets consist of the numbers (letters) $1, \ldots, a$. (It is easy to admit the possibility that one alphabet may have fewer symbols than the other. In that case a is the larger number of symbols, and the modifications needed in the theory below will be trivially obvious.) "Noisy" means that the letter sent may be garbled in transmission by noise (= chance error). Let the probability that the letter (number) j will be received when the letter i is sent be $w(j \mid i)$. Of course

$$w(1 \mid i) + \cdots + w(a \mid i) = 1, \quad i = 1, \ldots, a.$$

"Memoryless" means that the noise affects each letter independently of all other letters.

We shall assume that all the "words" we send have n letters. There is no reason for this assumption except that the theory has been developed under it, and that its elimination would make things more difficult. In ordinary writing, where the words of a language have different length, the word is determined by the spaces which separate it from its neighbors.

Such a "space" is really a letter of the alphabet. In a theory such as the one to be studied below, one seeks the most efficient use, in a certain sense, of the letters of the alphabet to transmit the words of a language. It is not at all certain that setting aside one letter of the alphabet to serve as a "space" is really the most efficient use of the letter (this even ignoring the fact that, because of noise, a "space" may not be received as a "space"). If one letter is not to serve to separate one word from the next then this will have to be done by the code (definition later), a requirement which would enormously complicate the problem. Thus we assume that all the words sent are sequences of n integers, each integer one of $1, \ldots, a$. (Any such sequence will be called an n-sequence.) Asymptotically (with n) this requirement will make no difference:

Suppose we have a dictionary of N words in a certain language. Here language could, for example, mean what is colloquially meant by a language, or it could mean the totality of words in a certain military vocabulary, or in a given diplomatic code book, or in a prescribed technical handbook. The words in the dictionary are of course what are colloquially known as words; they are written in the alphabet of the language (if the latter is English, for example, its alphabet is not the alphabet $1, \ldots, a$ of the letters sent over the channel) and they are not in general n-sequences or even of the same length (certainly not if the language is English). We wish to transmit (synonym for "send") any of these words in any arbitrary order and with any arbitrary frequency over the channel.

If there were no noise we would proceed as follows: Let n be the smallest integer such that $a^n \geqq N$. Construct in any manner a one-to-one correspondence between the words of the dictionary and N of the a^n n-sequences, this correspondence of course known to both sender and receiver. When one wishes to send any word in the dictionary one sends the n-sequence which corresponds to it. The receiver always receives the n-sequence correctly, and hence knows without error the word being sent.

When there is noise we must expect that in general there will be a positive probability that an n-sequence sent will be incorrectly received, i.e., received as another sequence (error of transmission). For example, when $w(j \mid i) > 0$ for every $i, j = 1, \ldots, a$, this will surely be the case. We would like then to be able to send over the channel any of the N words of the dictionary (more properly, any of the N n-sequences which correspond to the N words of the dictionary), with a probability $\leqq \lambda$, say, that any word of the dictionary (or rather n-sequence corresponding to it) sent will be incorrectly understood by the receiver. (Of course it is in general impossible that $\lambda = 0$. We shall be satisfied if $\lambda > 0$ is suitably small, and shall henceforth always assume, unless the contrary is explicitly stated, that $\lambda > 0$.) To achieve this we must proceed approximately as follows: Choose an n sufficiently large, for which $a^n > N$. There are of course a^n n-sequences. Establish a one-to-one correspondence between the N words of the dictionary and a properly selected set of N of the a^n sequences. The latter will be chosen so that the "distance" between any two of the selected sequences (in some sense yet to be established) is "sufficiently large" (the sense of this must also be established); it is to permit this that n must be sufficiently large. When one wishes to send any word of the dictionary one sends the n-sequence corresponding to it. The receiver operates according to the following rule: He always infers that the n-sequence actually sent is that one, of the N n-sequences which correspond to the N words of the dictionary, which is "nearest" to the n-sequence received ("resembles" it most). If the probability is $\geqq 1 - \lambda$ that, when any of the N n-sequences which correspond to words of the dictionary is sent, the n-sequence actually received is nearer to the one sent than to any of the other $N - 1$ n-sequences, the problem would be solved.

To summarize then, the problem is to choose N n-sequences in such a way that, whichever of the N n-sequences is sent, the receiver can correctly infer which one has been sent with probability $\geqq 1 - \lambda$. The solution is, roughly, to choose n sufficiently large so that the N n-sequences can be embedded in the space of all a^n n-sequences with enough distance between any two that, when any one of the N n-sequences is sent, the set of all n-sequences nearer to it than to any of the other $N - 1$ n-sequences, has probability $\geqq 1 - \lambda$ of being received.

We would like to emphasize that we have used the words "distance" and "nearest" in a very vague sense, and that they should not be taken literally. More generally, the problem is to construct a "code", as follows: a) to choose N among the a^n possible transmitted n-sequences as the sequences $u_1, \ldots, u_N$ to correspond to the words of the dictionary b) to divide up the a^n possible received n-sequences into N disjoint sets $A_1, \ldots, A_N$ such that, when u_i is sent, the probability that the received sequence will lie in A_i is $\geqq 1 - \lambda$, $i = 1, \ldots, N$. Then, when the receiver receives a sequence in A_i he decides that u_i was sent.

We shall now show that a solution can actually be effected with very little trouble. To simplify the discussion we shall assume that $a = 2$ (the general result will be obvious anyhow) and that $w(1 \mid 1) \neq w(1 \mid 2)$ (hence $w(2 \mid 1) \neq w(2 \mid 2)$); the explanation for this assumption will be given later in order not to cause a digression here. Let k be any integer, say the smallest, such that $2^k \geqq N$. Set up a one-to-one correspondence between the N words of the dictionary and any N k-sequences in any manner; henceforth we identify the words with these k-sequences. In order to transmit any k-sequence so that the probability of its correct reception is $\geqq 1-\lambda$ we simply send the same sequence m times consecutively, where m is a suitable integer which depends upon k and λ (and of course $w(\cdot \mid \cdot)$) and whose existence will shortly be demonstrated. Thus $n = km$. The idea of the method by which the receiver decides upon which sequence has been sent may be described simply as follows: The receiver knows that a letter b (unknown to him) was sent (not consecutively) m times and received as $b_1, \ldots, b_m$, say. He decides that

$$b = 1 \text{ if } \prod_{i=1}^{m} w(b_i \mid 1) > \prod_{i=1}^{m} w(b_i \mid 2),$$

$$b = 2 \text{ if } \prod_{i=1}^{m} w(b_i \mid 1) < \prod_{i=1}^{m} w(b_i \mid 2).$$

(In statistics this is called the maximum likelihood method.) If the above products are equal he may make any decision he pleases. We shall now prove that one can choose m so that, if the letter c is sent ($c = 1, 2$), then, calling the other letter c', the probability of all points $b = (b_1, \ldots, b_m)$ such that

$$\prod_{i=1}^{m} w(b_i \mid c) > \prod_{i=1}^{m} w(b_i \mid c')$$

is greater than $(1-\lambda)^{\frac{1}{k}}$. Thus the probability that all k letters are correctly received (inferred) is $> 1-\lambda$. (It would actually be better to apply the maximum likelihood method to the entire k-sequence than to each letter separately, but in an introduction such as this there is no need to dwell on this point.)

First we note that, for $c, c' = 1, 2$, and $c \neq c'$,

$$h(c) = \sum_{i=1}^{2} w(i \mid c) \log w(i \mid c) - \sum_{i=1}^{2} w(i \mid c) \log w(i \mid c') > 0.$$

To prove this maximize

$$\sum_{i=1}^{2} w(i \mid c) \log \pi_i$$

with respect to π_1 and π_2, subject to the restriction $\pi_1 + \pi_2 = 1$. The usual Lagrange multiplier argument gives that a *unique* maximum exists at $\pi_i = w(i \mid c)$, $i = 1, 2$. Since, by our assumption, $w(i \mid c) \neq w(i \mid c')$, $i = 1, 2$, the desired result follows.

Now let $h = \min(h(1), h(2)) > 0$. Let $Z_1, \ldots, Z_m$ be independent chance variables with the common distribution function

$$P\{Z_1 = i\} = w(i \mid c), \quad i = 1, 2$$

where c is 1 or 2. Let $c' \neq c$ be 1 or 2. From the preceding paragraph we have[1]

$$E\,[\log w(Z_1 \mid c) - \log w(Z_1 \mid c')] \geqq h > 0.$$

Hence, for $m > m(c)$ sufficiently large we have

$$P\left\{\prod_{i=1}^{m} w(Z_i \mid c) > e^{\left(\frac{mh}{2}\right)} \cdot \prod_{i=1}^{m} w(Z_i \mid c')\right\} > (1-\lambda)^{\frac{1}{k}}$$

by the law of large numbers. Now let $m = \max(m(1), m(2))$, and the desired result is evident.

We now digress for a moment to discuss the reason for the assumption that $w(1 \mid 1) \neq w(1 \mid 2)$. If $w(1 \mid 1) = w(1 \mid 2)$ and hence $w(2 \mid 1) = w(2 \mid 2)$, the distribution of the received letter is the same no matter which letter is sent. Thus it is impossible to infer from the letter received what letter has been sent; for any k there is essentially only one k-sequence which can be sent. Such a channel is clearly trivial and of no interest.

Returning to the problem under consideration, we have seen that, for $a = 2$, and this could also be shown generally, a trivial solution exists. To make the problem interesting and more realistic we ask how small an n will suffice to enable us to transmit N words with probability of error $\leqq \lambda$. This problem and two equivalent alternative versions may be listed as follows for future reference:

Form I). Given N and λ, how small an n will suffice?

Form II). Given n and λ, how big an N can be achieved?

Form III). Given n and N, how small a λ can be achieved?

A companion problem to the above problem (call it the first) is the (second) problem of actually constructing a code which would "implement" the answer to the first problem. In fact, it might a priori be reasonably thought that the first problem could not be solved without the second. This is not the case, and at present writing our knowledge about the first problem considerably exceeds our knowledge about the second problem. What success has been attained in the study of the first problem is due to certain brilliant fundamental work of Shannon, and subsequent work of the latter and other writers. This monograph is devoted to a discussion of some of the most outstanding results on the first problem for large n. We will not be restricted to the discrete channel without memory, and will consider other channels as well.

---

[1] $E\{\}$ denotes the expected value of the chance variable in brackets. $P\{\}$ denotes the probability of the set in brackets.

2. Combinatorial Preliminaries

2.1. Generated sequences. In this section we shall obtain certain combinatorial properties of sequences of n integers, each integer one of $1, \ldots, a$. The motivation for our interest in these properties will become apparent in Chapter 3. By at once proving the necessary facts we gain in efficiency at the expense of a temporary lack of motivation. To avoid the trivial we assume, throughout this monograph, that $a \geqq 2$ and $n \geqq 2$; the main interest in application will usually be for large n.

The use of combinatorial arguments is frequent in probability theory. We shall reverse this usual procedure and use formally probabilistic arguments to obtain combinatorial results. However, the probabilistic arguments to be employed are very simple and of no depth, and could easily be replaced by suitable combinatorial arguments. Their chief role is therefore one of convenience, to enable us to proceed with speed and dispatch.

The form of these arguments will always be this: It is required to give an upper or lower bound on the number of sequences in a certain set. A suitable probability distribution gives a lower bound α (upper bound β) to the probability of each of the sequences in the set, and a lower bound α_1 (upper bound β_1) to the probability of the set itself. Then $\frac{\alpha_1}{\beta}$ is a lower bound on the number of sequences in the set, and $\frac{\beta_1}{\alpha}$ is an upper bound on the number of sequences in the set. Of course, the proper choice of distribution is essential to obtaining the best values for these bounds.

A sequence of n integers, each one of $1, \ldots, a$, will be called an n-sequence. Let u_0 be any n-sequence. We define $N(i \mid u_0)$, $i = 1, \ldots, a$, as the number of elements i in u_0. Let u_0 and v_0 be two n-sequences. We define $N(i, j \mid u_0, v_0)$, $i, j = 1, \ldots, a$, as the number of k, $k = 1, \ldots, n$, such that the kth element of u_0 is i and the kth element of v_0 is j.

Let $\pi = (\pi_1, \ldots, \pi_a)$ be a vector with a non-negative components which add to one. The symbol π will always be reserved for such a vector, which will be called a π-vector, a (probability) distribution or a probability vector. When we wish to specify the number of components we shall speak of a π a-vector or a probability a-vector or a distribution on a points.

An n-sequence u_0 will be called a π-sequence or a πn-sequence if

$$|N(i \mid u_0) - n\pi_i| \leqq 2\sqrt{a n \pi_i (1 - \pi_i)}, \quad i = 1, \ldots, a. \tag{2.1.1}$$

Let $w(j \mid i)$, $i, j = 1, \ldots, a$, be any function of (i, j) such that

$$w(\cdot \mid i) = (w(1 \mid i), \ldots, w(a \mid i)), \quad i = 1, \ldots, a \tag{2.1.2}$$

is a probability vector. The significance of $w(\cdot \mid \cdot)$ will become apparent in Chapter 3; it will be called a "channel probability function" (c.p.f.). Let $\delta > 2a$ be a number to be determined later. An n-sequence v_0 will be said to be generated by an n-sequence u_0 if

$$\begin{aligned} &|N(i, j \mid u_0, v_0) - N(i \mid u_0)\, w(j \mid i)| \\ &\leqq \delta[N(i \mid u_0)\, w(j \mid i)\, (1 - w(j \mid i))]^{\frac{1}{2}} \end{aligned} \tag{2.1.3}$$

for all $i, j = 1, \ldots, a$.

Due to a convention of no importance but hallowed by tradition, all the logarithms in this monograph will be to the base 2. In order to avoid circumlocutions we shall adopt the convention that a quantity which, when taken literally, appears to be $0 \log 0$, is always equal to zero.

We define the function H of a π-vector π as follows:

$$H(\pi) = -\Sigma\, \pi_i \log \pi_i\,, \tag{2.1.4}$$

$H(\pi)$ is called the "entropy" of π. Its combinatorial importance, as we shall shortly see, is that the principal parts (for large n) of the logarithms of the numbers of n-sequences with certain properties are the entropies of certain distributions. In fact, we intend now to estimate the number $B_1(w \mid u_0)$ of n-sequences generated by u_0, where u_0 is any π-sequence, and the number $B_2(w \mid \pi)$ of different n-sequences generated by all π-sequences.

The function defined by (2.1.4) will often occur below, and should therefore, for convenience and brevity, have a name. However, we shall draw no implicit conclusions from its name, and shall use only such properties of H as we shall explicitly prove. In particular, we shall not erect any philosophical systems on H as a foundation. One reason for this is that we shall not erect any philosophical systems at all, and shall confine ourselves to the proof of mathematical theorems.

Let

$$(X_1, Y_1), \ldots, (X_n, Y_n) \tag{2.1.5}$$

be independent 2-vector chance variables with a common distribution determined by:

$$P\{X_1 = i\} = \pi_i, \quad i = 1, \ldots, a \tag{2.1.6}$$

$$P\{Y_1 = j \mid X_1 = i\} = w(j \mid i), \quad i, j = 1, \ldots, a. \tag{2.1.7}$$

Hence

$$P\{Y_1 = j\} = \sum_i \pi_i\, w(j \mid i) = \pi'_j \text{ (say)}, \quad j = 1, \ldots, a. \tag{2.1.8}$$

The vector

$$\pi' = (\pi'_1, \ldots, \pi'_a) \tag{2.1.9}$$

is of course a probability vector. Also

$$P\{X_1 = j \mid Y_1 = i\} = \frac{\pi_j w(i \mid j)}{\sum_k \pi_k w(i \mid k)} = w'(j \mid i) \text{ (say)}, \quad i, j = 1, \ldots, a. \tag{2.1.10}$$

Define the probability vectors

$$w'(\cdot \mid i) = (w'(1 \mid i), \ldots, w'(a \mid i)), \quad i = 1, \ldots, a. \tag{2.1.11}$$

Let

$$X = (X_1, \ldots, X_n) \tag{2.1.12}$$

and

$$Y = (Y_1, \ldots, Y_n). \tag{2.1.13}$$

Then X and Y are chance variables whose values are n-sequences.

First we note for later reference the trivial lemmas:

Lemma 2.1.1. We have

$$P\{X \text{ is a } \pi\text{-sequence}\} \geqq \frac{3}{4}. \tag{2.1.14}$$

This follows at once from (2.1.1) and CHEBYSHEV's inequality.

Lemma 2.1.2. Let u_0 be any n-sequence. Then

$$P\{Y \text{ is generated by } u_0 \mid X = u_0\} \geqq 1 - \varepsilon', \tag{2.1.15}$$

where $\varepsilon' \leqq \frac{a^2}{\delta^2} < \frac{1}{4}$, so that $\varepsilon' \to 0$ as $\delta \to \infty$.

This follows at once from (2.1.3) and CHEBYSHEV's inequality.

Let v_0 be any n-sequence generated by any π-sequence. Then an upper bound on $N(i \mid v_0)$ is

$$\begin{aligned} & n\pi_i' + \sum_j 2\sqrt{a n \pi_j}\, w(i \mid j) + \delta \sum_j \sqrt{n\pi_j + 2\sqrt{a n \pi_j}} \cdot \sqrt{w(i \mid j)} \\ & < n\pi_i' + 2\sqrt{a n} \sum_j \sqrt[4]{\pi_j}\, w(i \mid j) + 2a\delta\sqrt{n} \sum_j \sqrt[4]{\pi_j\, w(i \mid j)} \\ & < n\pi_i' + 2a^2\sqrt{n}\,(1+\delta)\sqrt[4]{\pi_i'} = V_{i1}. \end{aligned} \tag{2.1.16}$$

Similarly

$$V_{i0} = n\pi_i' - 2a^2\sqrt{n}\,(1+\delta)\sqrt[4]{\pi_i'}$$

is a lower bound on $N(i \mid v_0)$. Thus

$$\prod_i (\pi_i')^{V_{i1}} < P\{Y = v_0\} < \prod_i (\pi_i')^{V_{i0}}, \tag{2.1.17}$$

from (2.1.8) and the fact that the Y_i's are independently and identically distributed.

Since $-x \log x$ is bounded for $0 \leqq x \leqq 1$, so is

$$-\sqrt[4]{x} \log x = -4\sqrt[4]{x} \log \sqrt[4]{x}.$$

Hence
$$-\sum_i \sqrt[4]{\pi_i'}\log \pi_i'$$
is bounded above by a constant multiple of a.

Thus we have proved.

Lemma 2.1.3. Let v_0 be any n-sequence generated by any π-sequence. Then
$$\begin{aligned}&\exp_2\left\{-nH(\pi') - K_1 a^3(1+\delta)\sqrt{n}\right\} < P\{Y = v_0\}\\ &< \exp_2\left\{-nH(\pi') + K_1 a^3(1+\delta)\sqrt{n}\right\}\end{aligned} \tag{2.1.18}$$
where $K_1 > 0$ does not depend on v_0, a, n, π, δ, or $w(\cdot\mid\cdot)$.

From Lemma 2.1.3 we have at once
$$B_2(w\mid\pi) < \exp_2\left\{nH(\pi') + K_1 a^3(1+\delta)\sqrt{n}\right\}. \tag{2.1.19}$$
From Lemmas 2.1.1 and 2.1.2 we have
$$\begin{aligned}P\{Y \text{ is a sequence generated by a } \pi\text{-sequence}\} &\geqq \frac{3}{4}(1-\varepsilon')\\ &\geqq \frac{9}{16}.\end{aligned} \tag{2.1.20}$$
From (2.1.20) and Lemma 2.1.3 we have
$$B_2(w\mid\pi) > \frac{9}{16}\exp_2\left\{nH(\pi') - K_1 a^3(1+\delta)\sqrt{n}\right\}. \tag{2.1.21}$$
Thus we have proved

Lemma 2.1.4. We have
$$\begin{aligned}&\exp_2\left\{nH(\pi') - K_2 a^3(1+\delta)\sqrt{n}\right\} < B_2(w\mid\pi)\\ &< \exp_2\left\{nH(\pi') + K_2 a^3(1+\delta)\sqrt{n}\right\},\end{aligned} \tag{2.1.22}$$
where $K_2 = K_1 - \log\frac{9}{16} > 0$ and does not depend on a, n, π, δ, or $w(\cdot\mid\cdot)$.

Let v_0 be any n-sequence generated by the π-sequence u_0. Then
$$P\{Y = v_0 \mid X = u_0\} = \prod_{i,j} w(j\mid i)^{N(i,j\mid u_0,v_0)} \tag{2.1.23}$$
$$\begin{aligned}&\exp_2\left\{n\sum_{i,j}\pi_i w(j\mid i)\log w(j\mid i) + \sqrt{n}\,(2a+\delta)\sum_{i,j}\sqrt{w(j\mid i)}\log w(j\mid i)\right\}\\ &< P\{Y = v_0 \mid X = u_0\}\\ &< \exp_2\left\{n\sum_{i,j}\pi_i w(j\mid i)\log w(j\mid i) - \sqrt{n}\,(2a+\delta)\sum_{i,j}\sqrt{w(j\mid i)}\log w(j\mid i)\right\}.\end{aligned} \tag{2.1.24}$$
Thus we have

Lemma 2.1.5. Let u_0 be any πn-sequence, and v_0 be any n-sequence generated by u_0. Then
$$\begin{aligned}&\exp_2\left\{-n\sum_i \pi_i H(w(\cdot\mid i)) - \sqrt{n}\,(2a+\delta)a^2K_3\right\} < P\{Y = v_0\mid X = u_0\}\\ &< \exp_2\left\{-n\sum_i \pi_i H(w(\cdot\mid i)) + \sqrt{n}\,(2a+\delta)a^2K_3\right\},\end{aligned} \tag{2.1.25}$$
where $K_3 > 0$ does not depend on u_0, v_0, a, n, π, δ, or $w(\cdot\mid\cdot)$.

From Lemmas 2.1.2 and 2.1.5 we obtain

Lemma 2.1.6. Let u_0 be any πn-sequence. Then

$$\exp_2\left\{n\sum_i \pi_i H(w(\cdot\mid i)) - \sqrt{n}\,(2a+\delta)\,a^2K_4\right\} < B_1(w\mid u_0)$$
$$< \exp_2\left\{n\sum_i \pi_i H(w(\cdot\mid i)) + \sqrt{n}\,(2a+\delta)\,a^2K_4\right\}, \qquad (2.1.26)$$

where $K_4 = K_3 - \log\frac{3}{4} > 0$ and does not depend on u_0, a, n, π, δ, or $w(\cdot\mid\cdot)$.

From Lemmas 2.1.4 and 2.1.6 it follows that

$$H(\pi') - \sum_i \pi_i H(w(\cdot\mid i)) \geqq 0. \qquad (2.1.27)$$

Exactly as in Lemmas 2.1.4 and 2.1.6 one can prove

Lemma 2.1.7. The number $B_3(\pi)$ of πn-sequences satisfies

$$\exp_2\left\{nH(\pi) - K_5 a^{\frac{3}{2}}\sqrt{n}\right\} < B_3(\pi) < \exp_2\left\{nH(\pi) + K_5 a^{\frac{3}{2}}\sqrt{n}\right\}, \qquad (2.1.28)$$

where $K_5 > 0$ does not depend on a, n, π, δ, or $w(\cdot\mid\cdot)$.

2.2. Properties of the entropy function. If the components of π did not have to sum to unity, $H(\pi)$ would obviously be a strictly concave function of these components. It remains a strictly concave function of these components in the sub-space where they sum to unity.

Let W be the $(a\times a)$ matrix with element $w(j\mid i)$ in the jth row and i^{th} column. Writing π and π' as column vectors we have

$$\pi' = W\pi.$$

From this and the definition of concavity we have that $H(\pi') = H(W\pi)$ is a concave function of the components of π. If W is non-singular the concavity must be strict.

The function of π defined by

$$H(\pi') - \sum_i \pi_i H(w(\cdot\mid i)) = H(W\pi) - \sum_i \pi_i H(w(\cdot\mid i)) \qquad (2.2.1)$$

is the sum of a concave function of the components of π (i.e., $H(\pi')$) and a linear function of the components of π (i.e., $-\sum_i \pi_i H(w(\cdot\mid i))$), and is therefore a concave function of π. If W is non-singular the function (2.2.1) is strictly concave, and hence there is then a unique value of π at which the function attains its maximum. Even if W is singular the set of points at which the function (2.2.1) attains its maximum is a convex set. We shall always designate an arbitrary member of this set by $\bar{\pi}$.

All the chance variables of this section which we shall now employ to obtain properties of H are to be taken, without further statement, as discrete chance variables, each of whose components takes at most a values; without loss of generality the latter will be assumed to be $1, \ldots, a$. Thus X_1 of Section 2.1 is the most general such one-dimensional chance variable, and (X_1, Y_1) is the most general such two-dimensional chance variable. We shall not always indicate the dimensionality (always finite) of a chance variable. Thus Z will in general be a vector chance variable.

If Z is a (not necessarily one-dimensional) chance variable, the chance variable $P\{Z\}$ which is a function of Z will be defined as follows: When $Z = z$ then $P\{Z\} = P\{Z = z\}$. We define

$$H(Z) = -E \log P\{Z\}.$$

Hence

$$H(X_1) = H(\pi), \quad H(Y_1) = H(\pi').$$

We see that the entropy (function H) of a chance variable is simply the entropy of its distribution. It is nevertheless convenient thus to define H on the space of chance variables, because it makes for conciseness and brevity.

Now let $Z = (Z_1, Z_2)$ be a chance variable (Z_1 and Z_2 may be vectors) and define with probability one (w. p. 1) the (chance variable) function $P\{Z_2 \mid Z_1\}$ of Z as follows: When $(Z_1, Z_2) = (z_1, z_2)$ and $P\{Z_1 = z_1\} > 0$, then $P\{Z_2 \mid Z_1\} = P\{Z_2 = z_2 \mid Z_1 = z_1\}$. Also define

$$H(Z_2 \mid Z_1) = -E \log P\{Z_2 \mid Z_1\}.$$

Thus

$$H(Y_1 \mid X_1) = \sum_i \pi_i H(w(\cdot \mid i))$$

$$H(X_1 \mid Y_1) = \sum_i \pi'_i H(w'(\cdot \mid i)).$$

Finally, define the chance variable $P\{Z_2 \mid Z_1 = z_1\}$ when $P\{Z_1 = z_1\} > 0$, in the obvious manner as follows: When $Z_2 = z_2$ then $P\{Z_2 \mid Z_1 = z_1\} = P\{Z_2 = z_2 \mid Z_1 = z_1\}$. As before, we write

$$H(Z_2 \mid Z_1 = z_1) = -E \log P\{Z_2 \mid Z_1 = z_1\}.$$

Thus

$$H(Z_2 \mid Z_1) = \sum_i H(Z_2 \mid Z_1 = i)\, P\{Z_1 = i\}.$$

Since, w. p. 1,

$$\log P\{Z_1\} + \log P\{Z_2 \mid Z_1\} = \log P\{Z_1, Z_2\} = \log P\{Z_2\} + \log P\{Z_1 \mid Z_2\}$$

we have

$$H(Z_1, Z_2) = H(Z_1) + H(Z_2 \mid Z_1) = H(Z_2) + H(Z_1 \mid Z_2). \quad (2.2.2)$$

In particular

$$H(\pi) - \sum_i \pi'_i H(w'(\cdot \mid i)) = H(\pi') - \sum_i \pi_i H(w(\cdot \mid i)). \quad (2.2.3)$$

An elementary argument with a Lagrange multiplier shows that

$$H(\pi) \leqq \log a, \text{ with a unique maximum at}$$

$$\pi = \left(\frac{1}{a}, \ldots, \frac{1}{a}\right). \tag{2.2.4}$$

We now find the maximum of $H(\pi)$, subject to the constraint $\pi_1 \leqq a'$. Fix π_1. The same argument with a Lagrange multiplier gives

$$H(\pi) \leqq -\pi_1 \log \pi_1 - (1-\pi_1) \log \left(\frac{1-\pi_1}{a-1}\right),$$

$$\text{with a unique maximum at} \tag{2.2.5}$$

$$\pi = \left(\pi_1, \frac{1-\pi_1}{a-1}, \ldots, \frac{1-\pi_1}{a-1}\right).$$

Now vary π_1, subject to $\pi_1 \leqq a'$. We obtain:

$$\text{if } a' < \frac{1}{a}, \quad H(\pi) \leqq -a' \log a' - (1-a') \log (1-a') + (1-a') \log (a-1) \tag{2.2.6}$$

$$\text{with a unique maximum at}$$

$$\pi = \left(a', \frac{1-a'}{a-1}, \ldots, \frac{1-a'}{a-1}\right)$$

$$\text{if } a' \geqq \frac{1}{a}, \text{ then (2.2.4) holds.} \tag{2.2.7}$$

Write

$$\pi_{ij} = P\{X_1 = i, Y_1 = j\}.$$

Now maximize

$$H(\{\pi_{ij}\}) = -\sum_{i,j} \pi_{ij} \log \pi_{ij},$$

subject to the constraints

$$\sum_j \pi_{ij} = \pi_i, \quad i = 1, \ldots, a$$

$$\sum_i \pi_{ij} = \pi'_j, \quad j = 1, \ldots, a.$$

An elementary argument with Lagrange multipliers shows that there is a unique maximum when

$$\pi_{ij} = \pi_i \cdot \pi'_j, \quad i, j = 1, \ldots, a.$$

Hence we have that

$$H(X_1, Y_1) \leqq H(X_1) + H(Y_1) \tag{2.2.8}$$

with equality when and only when the following *equivalent* conditions hold:

$$X_1 \text{ and } Y_1 \text{ are independent.} \tag{2.2.9}$$

$$w(\cdot \mid i) \text{ is the same for all } i \text{ such that } \pi_i > 0. \tag{2.2.10}$$

$$w'(\cdot \mid i) \text{ is the same for all } i \text{ such that } \pi'_i > 0. \tag{2.2.11}$$

From (2.2.2) and (2.2.8) we obtain

$$H(X_1 \mid Y_1) \leqq H(X_1) \tag{2.2.12}$$

$$H(Y_1 \mid X_1) \leqq H(Y_1). \tag{2.2.13}$$

(Of course there is no need to write both (2.2.12) and (2.2.13); one implies the other by interchange of the roles of X_1 and Y_1, since (X_1, Y_1) is the most general representation in two dimensions of the chance variables considered in this section. This may be done in other formulae of the present section where we have not always bothered to state the dual result. Incidentally, (2.2.13) is simply (2.1.27).) The equality sign in (2.2.12) and (2.2.13) holds when and only when (2.2.9), (2.2.10), or (2.2.11) holds.

Let $(Z_1, \ldots, Z_k)$ be any chance variable. By repeated application of (2.2.8) we obtain

$$H(Z_1, \ldots, Z_k) \leqq H(Z_1) + \cdots + H(Z_k) \tag{2.2.14}$$

with equality when and only when $Z_1, Z_2, \ldots, Z_k$ are independent. Now apply (2.2.12) with $X_1 = Z_1$, $Y_1 = Z_3$, and Z_2 fixed, and then take the expected value with respect to Z_2. One obtains

$$H(Z_1 \mid Z_2, Z_3) \leqq H(Z_1 \mid Z_2),$$

with the equality sign holding if, and only if, Z_1 and Z_3 are independently distributed whenever Z_2 is fixed at any point z_2 such that $P\{Z_2 = z_2\} > 0$. Application of this procedure, with Z_2 replaced by $Z_2, \ldots, Z_{k-1}$ and Z_3 by Z_k, gives

$$H(Z_1 \mid Z_2, \ldots, Z_k) \leqq H(Z_1 \mid Z_2, \ldots, Z_{k-1}) \tag{2.2.15}$$

with the equality sign holding if, and only if, Z_1 and Z_k are independently distributed whenever $Z_2, \ldots, Z_{k-1}$ are fixed at any point $(z_2, \ldots, z_{k-1})$ such that $P\{(Z_2, \ldots, Z_{k-1}) = (z_2, \ldots, z_{k-1})\} > 0$.

Since $x \log x \neq 0$ except when $x = 0$ or 1, we have that $H(\pi) = 0$ when, and only when, all components of π, with one exception, are zero. We note also that $H(X_1 \mid Y_1) = 0$ implies that, for every i such that $P\{Y_1 = i\} > 0$, all the components of the vector $w'(\cdot \mid i)$, with one exception, are zero. (Of course the dual statement about $H(Y_1 \mid X_1)$ and $w(\cdot \mid i)$ holds.) Also, $H(X_1 \mid Y_1)$ and $H(Y_1 \mid X_1)$ need not be zero at the same time, as the following example shows: $a = 4$, $\pi_3 = \pi_4 = 0$, $\pi_1 > 0$, $\pi_2 > 0$; $0 < w(1 \mid 1) < 1$, $0 < w(3 \mid 2) < 1$, $w(1 \mid 1) + w(2 \mid 1) = 1$, $w(3 \mid 2) + w(4 \mid 2) = 1$. Then $H(Y_1 \mid X_1) > 0$, $H(X_1 \mid Y_1) = 0$.

Let ϕ be any function on the integers $1, \ldots, a$. Then

$$H(Z_1 \mid Z_2) = H(Z_1 \mid Z_2, \phi(Z_2)) \leqq H(Z_1 \mid \phi(Z_2)) \tag{2.2.16}$$

by (2.2.15). Let x_1 and x_2 be positive numbers. Then

$$-(x_1 + x_2)^{x_1+x_2} < -x_1^{x_1} x_2^{x_2}.$$

Hence

$$H(Z) \geqq H(\phi(Z)), \tag{2.2.17}$$

with the equality sign holding if, and only if, $\phi(Z)$ has a single-valued inverse with probability one. The inequality (2.2.17) also follows from the concavity of H and JENSEN's inequality (which anyhow can be proved by essentially the argument we gave).

Remarks

The definition of entropy was given by SHANNON [*1*] and WIENER [*1*]. (References are to the bibliography at the end of the monograph.) SHANNON (ibid.) conjectured and later, in 1957, proved (see Remarks on Chapter 3) Theorem 3.2.1 below, in which the entropy function plays a crucial role.

SHANNON pointed out [*1*] that the function H is, to within a multiplicative constant, uniquely determined by the following conditions: 1) $H(\pi)$ is a continuous function of $\pi_1, \ldots, \pi_a$. 2) $H\left(\frac{1}{a}, \ldots, \frac{1}{a}\right)$ is monotonically increasing with a. 3) If $\pi_a = q_1 + q_2$, then

$$H(\pi_1, \ldots, \pi_{a-1}, q_1, q_2) = H(\pi_1, \ldots, \pi_a) + \pi_a H\left(\frac{q_1}{\pi_a}, \frac{q_2}{\pi_a}\right).$$

However, he also remarked that he would not make use of these properties, and that the reason for the introduction of the entropy function is its role in coding theory.

The various properties of H described in Section 2.2 were either pointed out by SHANNON or else are obvious.

The introduction of generated sequences and the lemmas of Section 2.1 are due to the author [*1*]. It seems to the author that the lemmas of Section 2.1 also give an intuitive explanation of the meaning and significance of H. However, what is intuitive is a purely subjective matter on which opinions may easily differ widely.

The coefficients of $\sqrt{n}$ in many of the lemmas of Section 2.1 could easily be improved. The given coefficients suffice for our purposes.

3. The Discrete Memoryless Channel

3.1. Description of the channel. Let $w(j \mid i)$, $i, j = 1, \ldots, a$, be non-negative numbers such that

$$w(\cdot \mid i) = \big(w(1 \mid i), \ldots, w(a \mid i)\big)$$

is a probability a-vector. The function $w(\cdot \mid \cdot)$ will now play the role of a "channel probability function" (c.p.f.).

Our channel sends (transmits) n-sequences and receives n-sequences. The received n-sequence is a chance sequence (all of its components are chance variables) whose distribution depends (solely) upon the sequence sent in a manner now to be described. Let $u_0 = (x_1, \ldots, x_n)$ be the sequence sent, and

$$v(u_0) = \bigl(Y_1(u_0), \ldots, Y_n(u_0)\bigr) \tag{3.1.1}$$

be the chance sequence received. The $Y_i(u_0)$, $i = 1, \ldots, n$, are independent chance variables with distributions given by

$$P\{Y_i(u_0) = j\} = w(j \mid x_i), \quad i = 1, \ldots, n \quad j = 1, \ldots, a. \tag{3.1.2}$$

Hence

$$P\{v(u_0) = (j_1, \ldots, j_n)\} = \prod_{i=1}^{n} w(j_i \mid x_i). \tag{3.1.3}$$

A system for transmitting sequences is called a channel. It is not necessary for our purposes to define it more precisely than that. What is essential is to define the distribution of $v(u_0)$ precisely, and this we shall always do. A channel which satisfies (3.1.2) and (3.1.3) will be called a discrete memoryless channel (d.m.c.).

We have assumed that the input alphabet (set of letters used for sending messages) and the output alphabet (set of letters used for receiving messages) are both $\{1, \ldots, a\}$. If one alphabet contains more symbols than the other the changes needed below will always be trivial and obvious. (See also remark 10 at the end of this chapter.) The actual symbols used are of no consequence; only the c.p.f. is relevant.

A code of length N is a system

$$\{(u_1, A_1), \ldots, (u_N, A_N)\} \tag{3.1.4}$$

where

$$u_i, \ i = 1, \ldots, N, \text{ is an } n\text{-sequence} \tag{3.1.5}$$

$$A_i, \ i = 1, \ldots, N, \text{ is a set of } n\text{-sequences} \tag{3.1.6}$$

$$A_1, \ldots, A_N \text{ are disjoint.} \tag{3.1.7}$$

The code is used as follows: The N sequences u_i are made to correspond in an arbitrary manner to the N words of a dictionary. When one wishes to send the ith word one sends u_i. When the (chance) received sequence lies in A_j the receiver concludes that u_j has been sent. Consequently it would seem sensible to have $A_1 \cup A_2 \cup \cdots \cup A_N$ be the entire space of n-sequences, but we do not insist on this. Nor do we insist that the sequences $u_1, \ldots, u_N$ all be different. If the sequence received does not lie in any of the A_j, the receiver may draw any conclusion he wishes about the word that has been sent.

The probability $\lambda(u_i)$ that the ith word (u_i) will be incorrectly decoded (understood) is of course given by

$$\lambda(u_i) = 1 - P\{v(u_i) \varepsilon A_i\}. \tag{3.1.8}$$

We define the "average probability of error" of the code as

$$\bar{\lambda} = \frac{1}{N} \sum_{i=1}^{N} \lambda(u_i) \tag{3.1.9}$$

and the "probability of error" of the code as

$$\lambda_M = \max_i \lambda(u_i). \tag{3.1.10}$$

A code of length N and probability of error $\leqq \lambda$ is a code of length N for which $\lambda_M \leqq \lambda$; we call such a code a code (n, N, λ) or a code (N, λ) for short.

We shall frequently prove below the existence of a code (N, λ) where N is not an integer. In such a case it is always to be understood that the code whose existence has been proved has length equal to the smallest integer $\geqq N$. This obvious convention is to apply throughout the monograph.

The following lemma is of utmost simplicity. It will not find application until a later chapter, but is given here to show how, from the existence of one kind of code one can infer the existence of another kind of code.

Lemma 3.1.1. If a code has length $2N$ and average probability of error $\leqq \lambda$, it contains a subcode (code which is a subset of the original code) of length N and probability of error $\leqq 2\lambda$.

Proof: If

$$\sum_i \lambda(u_i) \leqq 2N\lambda$$

then for at most N i's can $\lambda(u_i) \geqq 2\lambda$.

Let π be any probability a-vector, and π' a probability a-vector defined as in Chapter 2 as a function of π and $w(\cdot \mid \cdot)$. Define (see (2.2.1))

$$\begin{aligned} C &= \max_\pi \left(H(\pi') - \sum_i \pi_i H(w(\cdot \mid i))\right) \\ &= \max_\pi \left(H(\pi) - \sum_i \pi'_i H(w'(\cdot \mid i))\right). \end{aligned} \tag{3.1.11}$$

We write max instead of sup because the maximum is obviously attained, from continuity considerations. Let $\bar{\pi}$ be a vector π for which the maximum is attained, and $\bar{\pi}'$ be the corresponding vector π'.

When is $C = 0$? We have

$$\begin{aligned} H(\pi') - \sum_i \pi_i H(w(\cdot \mid i)) &= H(Y_1) - H(Y_1 \mid X_1) \\ &= H(X_1) - H(X_1 \mid Y_1) \\ &= H(X_1) + H(Y_1) - H(X_1, Y_1) \end{aligned} \tag{3.1.12}$$

by (2.2.2). If $C = 0$ then these are zero for every π. From (2.2.8) and (2.2.10) we conclude: $C = 0$ when and only when $w(\cdot \mid i)$ is the same for *all* i, $i = 1, \ldots, a$. When this is so the distribution of the received letter is the same, no matter what letter is sent. It is therefore impossible really to infer from the sequence received what sequence has been sent, no matter what code may be used. It follows easily that, in this case, $\frac{1}{1-\lambda}$ is an upper bound on the length of the code for every n.

From (2.2.4) it follows that $C \leqq \log a$. When is $C = \log a$? In that case

$$H\left(w(\cdot \mid i)\right) = 0, \quad i = 1, \ldots, a.$$

Hence, for each i there is a $j(i)$ such that

$$w\left(j(i) \mid i\right) = 1.$$

Moreover, if, for $i_1 \neq i_2$, $j(i_1) = j(i_2)$, we could not have $\pi' = \left(\frac{1}{a}, \ldots, \frac{1}{a}\right)$ which is required by (2.2.4). We conclude that $C = \log a$ if, and only if, the matrix W of Section 2.2 can be obtained from the identity matrix by a permutation of the rows. But then the channel is "noiseless", i.e., each letter is transmitted without error.

The number C enters in an essential way into the statements of Theorems 3.2.1 and 3.3.1 below. A number which enters in this way into lower and upper bounds on the lengths of possible codes (n, N, λ), with $0 < \lambda < 1$, (lower bound for some codes and upper bound on *all* such codes) for any channel (the general channel will be defined later), is called the "capacity" of the channel[1]. Thus Theorems 3.2.1 and 3.3.1 could be approximately summarized by saying that together they prove that C is the capacity of the d.m.c. The reasons for the choice of the word "capacity" will become clear from the theorems.

3.2. A coding theorem. We now prove the following coding theorem:

Theorem 3.2.1. Let λ, $0 < \lambda \leqq 1$ be arbitrary. There exists a positive constant K such that, for any n, there exists a code (n, N, λ) with

$$N > \exp_2\left\{nC - K\sqrt{n}\right\}. \tag{3.2.1}$$

The number K will in general depend on λ and a, but it can be chosen independently of the function $w(\cdot \mid \cdot)$.

Proof: Suppose $\delta > 2a$ is such that Lemma 2.1.2 holds with $\varepsilon' \leqq \frac{\lambda}{2}$. Let (3.1.4) be a code (n, N, λ) which fulfills the following conditions:

u_i, $i = 1, \ldots, N$, is a $\bar{\pi}$-sequence (3.2.2)

A_i, $i = 1, \ldots, N$, consists of all n-sequences (3.2.3)
generated by u_i and not in $A_1 \cup \cdots \cup A_{i-1}$.

[1] For a precise definition of capacity see Section **7.7**

it is impossible to prolong the code (3.1.4) by adding an element (u_{N+1}, A_{N+1}) such that the resulting set is a code $(n, N+1, \lambda)$ and (u_{N+1}, A_{N+1}) satisfies (3.2.2) and (3.2.3) for $i = N+1$. (3.2.4)

That such codes (of length at least one) exist follows from Lemma 2.1.2. We shall now show that any code with the above properties has the length described in the theorem.

Let u_0 be any $\bar{\pi}$-sequence not in the set $\{u_1, \ldots, u_N\}$. Then

$$P\{v(u_0) \text{ is generated by } u_0 \text{ and belongs to } A_1 \cup A_2 \cup \cdots \cup A_N\} > \frac{\lambda}{2}. \tag{3.2.5}$$

For, if this were not so, we could prolong the code by adding (u_0, A), where A is the totality of n-sequences generated by u_0 and not in $A_1 \cup \cdots \cup A_N$.

It follows from Lemma 2.1.2 and (3.2.3) that (3.2.5) holds for every $\bar{\pi}$-sequence, i.e., even when u_0 is in the set $\{u_1, \ldots, u_N\}$. Let $(X_1, Y_1), \ldots, (X_n, Y_n)$ be the chance variables defined in (2.1.5) for $\pi = \bar{\pi}$. From Lemma 2.1.1 and (3.2.5) we have

$$P\{Y \varepsilon A_1 \cup \cdots \cup A_N\} > \frac{3}{4} \cdot \frac{\lambda}{2}. \tag{3.2.6}$$

From (3.2.6) and Lemma 2.1.3 it follows that $A_1 \cup \cdots \cup A_N$ contains at least

$$\frac{3\lambda}{8} \cdot \exp_2 \left\{ n H(\bar{\pi}') - K_1 a^3 (1 + \delta) \sqrt{n} \right\} \tag{3.2.7}$$

n-sequences. From Lemma 2.1.6 we have that $A_1 \cup \cdots \cup A_N$ contains at most

$$N \cdot \exp_2 \left\{ n \sum_i \bar{\pi}_i H(w(\cdot \mid i)) + \sqrt{n}\,(2a + \delta) a^2 K_4 \right\} \tag{3.2.8}$$

n-sequences. Hence

$$\begin{aligned} N &> \frac{3\lambda}{8} \cdot \exp_2 \left\{ nC - \sqrt{n}\,[K_1 a^3 (1+\delta) + K_4 a^2 (2a + \delta)] \right\} \\ &\geqq \exp_2 \left\{ nC - K \sqrt{n} \right\} \end{aligned} \tag{3.2.9}$$

with

$$K = K_1 a^3 (1 + \delta) + K_4 a^2 (2a + \delta) - \log \frac{3\lambda}{8}.$$

This proves the theorem.

Actually we have proved more. Suppose we construct a code with probability of error $\leqq \lambda$ by adding successive elements as follows: We begin with (u_1, A_1), then add (u_2, A_2), etc. At every step, say the ith, $i = 1, \ldots, N$, (3.2.2) and (3.2.3) are satisfied, and of course $\lambda(u_i) \leqq \lambda$.

Suppose also we try to add these elements in the most unfavorable way possible so as to make prolongation of the code as difficult as possible. In spite of this, when the code can no longer be prolonged its length N satisfies (3.2.1).

Suppose $\lambda < \frac{1}{2}$. If $\delta = a\sqrt{\frac{2}{\lambda}}$, the ε' of Lemma 2.1.2 is $\leqq \frac{\lambda}{2}$. Hence it suffices to take

$$K = K_1 a^3\left(1 + a\sqrt{\frac{2}{\lambda}}\right) + K_4 a^2\left(2a + a\sqrt{\frac{2}{\lambda}}\right) - \log\frac{3\lambda}{8}.$$

In general, this is much too strong a requirement, and a smaller K will suffice.

By modifying the above proof one can easily show that, for any $\varepsilon > 0$, there are two positive constants, c_1 and c_2, such that, for any n, there exists a code

$$(n, 2^{n(C-\varepsilon)}, c_1 e^{-nc_2}). \tag{3.2.10}$$

The result (3.2.10) is of the form III of Chapter 1. Theorem 3.2.1 is of the form II.

3.3. The strong converse. The strong converse of Theorem 3.2.1 states: Let $\varepsilon > 0$ and λ, $0 \leqq \lambda < 1$, be arbitrary. For n sufficiently large there does not exist a code $(n, 2^{n(C+\varepsilon)}, \lambda)$. (In Chapter 7 it will become clear why this is called the *strong* converse. It is now easy to see why C is called the capacity.) We will prove the somewhat stronger

Theorem 3.3.1. Let λ, $0 \leqq \lambda < 1$, be arbitrary. There exists a positive constant K' such that, for any n, there does not exist a code $(n, 2^{nC+K'\sqrt{n}}, \lambda)$. ($K'$ in general depends upon λ and a, but need not depend on $w(\cdot \mid \cdot)$.)

It is sufficient to prove the theorem for $\lambda > 0$, as it will then a fortiori be true when $\lambda = 0$. We therefore assume $\lambda > 0$ and begin by proving

Lemma 3.3.1. There exists a positive K_1' which depends on λ but not on n, a, π, δ, or $w(\cdot \mid \cdot)$, and which has the following property: Let (3.1.4) be any code (n, N, λ) such that, for $i = 1, \ldots, N$, u_i is a π-sequence and A_i contains only n-sequences generated by u_i. Then

$$N < \exp_2\left\{n\left[H(\pi') - \sum_i \pi_i H(w(\cdot \mid i))\right] + \sqrt{n}\, K_1' a^3 (2a + \delta)\right\}.$$

(A code which fulfills the conditions of the lemma will exist if δ is sufficiently large so that ε' of Lemma 2.1.2 is $\leqq \lambda$. For this purpose it always suffices for δ to be $a\sqrt{\frac{1}{\lambda}}$. However, δ must be $> 2a$ (see Section 2.1).)

Proof: From Lemma 2.1.5 we obtain that $A_i, i = 1, \ldots, N$, contains at least

$$(1-\lambda)\exp_2\left\{n\sum_i \pi_i H(w(\cdot\mid i)) - K_3 a^2(2a+\delta)\sqrt{n}\right\}$$

sequences, where $K_3 > 0$ is the constant of Lemma 2.1.5. Since the A_i are disjoint the set $A_1 \cup \cdots \cup A_N$ contains at least

$$N(1-\lambda)\exp_2\left\{n\sum_i \pi_i H(w(\cdot\mid i)) - K_3 a^2(2a+\delta)\sqrt{n}\right\}$$

sequences. From Lemma 2.1.4 it follows that $A_1 \cup \cdots \cup A_N$ contains fewer than

$$\exp_2\left\{nH(\pi') + K_2 a^3(1+\delta)\sqrt{n}\right\}$$

sequences, where K_2 is the constant of Lemma 2.1.4. Hence

$$N < (1-\lambda)^{-1}\times$$
$$\times\exp_2\left\{n\left[H(\pi') - \sum_i \pi_i H(w(\cdot\mid i))\right] + \sqrt{n}\left[K_2 a^3(1+\delta) + K_3 a^2(2a+\delta)\right]\right\}.$$

From this the desired result follows for $K_1' = K_2 + K_3 - \log(1-\lambda)$.

Lemma 3.3.2. Let δ be so large that ε' of Lemma 2.1.2 is $< \frac{1-\lambda}{2}$. (It always suffices for δ to be $a\sqrt{\frac{2}{1-\lambda}}$, provided the latter is $> 2a$.) Let (3.1.4) be any code (n, N, λ) such that, for $i = 1, \ldots, N$, u_i is a π-sequence. Then

$$N < \exp_2\left\{n\left[H(\pi') - \sum_i \pi_i H(w(\cdot\mid i))\right] + \sqrt{n}\,K_2' a^3(2a+\delta)\right\},$$

where $K_2' > 0$ depends on λ but not on n, a, π, δ, or $w(\cdot\mid\cdot)$.

Proof: From each A_i, $i = 1, \ldots, N$, delete those sequences not generated by u_i; call the result after deletion A_i'. Then

$$\{(u_1, A_1'), \ldots, (u_N, A_N')\}$$

is a code $\left(n, N, \frac{1+\lambda}{2}\right)$. This code satisfies the hypotheses of Lemma 3.3.1 with λ replaced by $\frac{1+\lambda}{2}$. Hence the conclusion of Lemma 3.3.1 holds.

Proof of Theorem 3.3.1. Any code (n, N, λ) can be divided into (not necessarily disjoint) subcodes such that the sequences u_i of any subcode are all π-sequences, for some π whose components are integral multiples of $\frac{1}{n}$. There are fewer than $(n+1)^a$ such subcodes. From Lemma 3.3.2 we obtain that

$$N < (n+1)^a \exp_2\left\{n C + K_2' a^3 (2a+\delta)\sqrt{n}\right\},$$

with δ as described in the lemma. This proves the theorem.

3.4. Strong converse for the binary symmetric channel. We now prove a result stronger than the general Theorem 3.3.1 for the special case of the binary symmetric channel. A d.m.c. is called binary symmetric if $a = 2$ and $w(1 \mid 1) = w(2 \mid 2)$ (hence $w(2 \mid 1) = w(1 \mid 2)$). For typographical simplicity write $q = w(1|1) = w(2|2)$, $p = w(1|2) = w(2|1)$. Then $\sum_i \pi_i H(w(\cdot \mid i)) = -p \log p - q \log q$ does not depend on π. Hence C would be achieved by a π for which $\pi' = \left(\frac{1}{2}, \frac{1}{2}\right)$. But $\pi = \left(\frac{1}{2}, \frac{1}{2}\right)$ is such a π. Hence

$$C = 1 + p \log p + q \log q.$$

We now prove the following theorem for a binary symmetric channel:

Theorem 3.4.1. Suppose $1 > q > \frac{1}{2}$ and $0 \leqq \lambda < \frac{1}{2}$. There exists a positive constant K^* such that, for n sufficiently large, there does not exist a code $(n, 2^{nC - K^*\sqrt{n}}, \lambda)$. (The constant K^* depends upon q and λ. To make it independent of q and λ one may set $K^* = 0$.)

Proof: Let u_0 be any n-sequence sent. Since $q > \frac{1}{2}$ the sequence u_0 has the greatest probability (q^n) of being received among all possible received sequences. Among the latter there are n sequences with the second largest probability $(q^{n-1} p)$, $\binom{n}{2}$ sequences with the third largest probability $(q^{n-2} p^2)$, etc. Let K'' be the smallest integer such that

$$\sum_{j=0}^{K''} \binom{n}{j} q^{n-j} p^j \geqq 1 - \lambda.$$

Then each A_i, $i = 1, \ldots, N$, of the code (n, N, λ) in (3.1.4) must contain at least

$$T = \sum_{j=0}^{K''} \binom{n}{j}$$

n-sequences. Since there are 2^n n-sequences we must have

$$N \leqq \frac{2^n}{T} < \frac{2^n}{\binom{n}{K''}}.$$

Define β_λ by

$$\frac{1}{\sqrt{2\pi}} \int_{-\infty}^{\beta_\lambda} \exp\left\{-\frac{1}{2} t^2\right\} dt = 1 - \lambda.$$

Since $\lambda < \frac{1}{2}$ we have $\beta_\lambda > 0$. From the DeMoivre-Laplace limit theorem (e.g., FELLER [1], p. 172) we have that, for n sufficiently large,

$$K'' > np + \frac{1}{2}\sqrt{npq}\, \beta_\lambda = np + B\sqrt{n}, \text{ say}.$$

We now employ STIRLING's formula

$$r^r \sqrt{2\pi r} \exp\{-r\} < r! < r^r \sqrt{2\pi r} \exp\left\{-r + \frac{1}{12r}\right\}$$

to obtain a lower bound on $\binom{n}{np + B\sqrt{n}}$. We obtain that, for large n,

$$-\log\binom{n}{K''} < n(p \log p + q \log q) - \sqrt{n}\; B \log\left(\frac{q}{p}\right) + c \log n,$$

where c is a positive constant. Hence

$$N < \exp_2\left\{n C - \sqrt{n}\; B \log\left(\frac{q}{p}\right) + c \log n\right\}$$

for n sufficiently large. This proves the theorem.

Theorem 3.4.1 could already have been obtained from the proof of Theorem 3.3.1 by a more careful examination of the constants involved. Since the latter depend continuously on $w(\cdot|\cdot)$, the following result (whose proof we leave to the reader) holds for binary "almost" symmetric channels:

Theorem 3.4.2. In a d.m.c. let $a = 2$, $w(1 \mid 1) > \frac{1}{2}$, and

$$|w(1 \mid 1) - w(2 \mid 2)|$$

be sufficiently small. Then, if $0 \leqq \lambda < \frac{1}{2}$ and n is sufficiently large, there does not exist a code $(n, 2^{nC - K_0\sqrt{n}}, \lambda)$, where K_0 is a positive constant which depends on $w(\cdot \mid \cdot)$ and λ.

3.5. The finite-state channel with state calculable by both sender and receiver. Suppose given t c.p.f.'s, $\{w(\cdot \mid \cdot \mid s), s = 1, \ldots, t\}$. Each c.p.f. is called a "state" of the channel. When the word (n-sequence) $u_0 = (x_1, \ldots, x_n)$ is about to be sent the channel is in some state, say c_1 (what this means we shall see in a moment), which is known to both sender and receiver. The sender sends the first letter, x_1, which is received as the chance variable $Y_1(u_0)$. The distribution of $Y_1(u_0)$ is $w(\cdot \mid x_1 \mid c_1)$. The channel now moves into a state $c_2 = \phi(c_1, x_1)$, where ϕ is a given function whose domain is the Cartesian product of the sets $\{1, \ldots, t\}$ and $\{1, \ldots, a\}$. The state c_2 is of course known to the sender. The state c_2 is also known to the receiver, either because there exists a function ϕ' defined on the same domain as ϕ and such that $\phi'(c_1, Y_1(u_0)) = c_2$, or else through some other mechanism which need not concern us. The sender now sends the next letter, x_2, which is received as the chance variable $Y_2(u_0)$. The conditional distribution of $Y_2(u_0)$, given u_0, c_1, and $Y_1(u_0)$, is $w(\cdot|x_2|c_2)$. This process is repeated for each letter. Thus, let the channel be in state c_i when the ith letter x_i is to be sent, $i = 1, \ldots, n$. The state c_i is known to both sender and receiver. The letter x_i is received as the chance variable $Y_i(u_0)$, whose conditional distribution, given u_0, c_1, $Y_1(u_0), \ldots, Y_{i-1}(u_0)$, is $w(\cdot \mid x_i \mid c_i)$. If $i < n$ the channel moves into state $c_{i+1} = \phi(c_i, x_i)$. The state c_{i+1} is also known to the receiver; if the mechanism through which it is known is the function ϕ' then $c_{i+1} = \phi'(c_i. Y_i(u_0))$.

Thus the state of the channel for any letter transmitted is determined, through the function ϕ, by the previous letter transmitted and the state for that letter, and is therefore "calculable" by the sender. If there is a function ϕ' it is also calculable by the receiver, and the title of this section is justified. However, the theorems and arguments of this section are valid whether or not there is a function ϕ'. On the other hand, it is essential for the validity of the argument of this section (and also of the next) that the state always be *determined* by the previous letter transmitted and the state for that letter.

In principle there need be no a priori restrictions on the function ϕ. The existence of a function ϕ' as the mechanism through which the receiver learns the state of the channel implies restrictions on ϕ, or on the w's, or on both. Further restrictions are imposed by the following assumption, which we make for the remainder of this section (and the next):

> Let d_1 and d_2 be any two states of the channel, distinct or not. There exists some integer $k(d_1, d_2)$ and a k-sequence $y_1, \ldots, y_k$ such that, if d_1 is the initial state of the channel and one sends over it, consecutively, the letters $y_1, \ldots, y_k$, the channel is in the state d_2 after y_k has been sent. (3.5.1)

Thus (3.5.1) says that it is possible to go from any state of the channel to any other state. While the number k of letters (above) depends upon the "path" chosen, it need never be greater than at.

It is a consequence of (3.5.1) that the stationary Markov chain with transition matrix M given in (3.5.8) below has no transient states and contains only one ergodic class (in the terminology of DOOB [*1*], p. 179) or only one closed set (in the terminology of FELLER [*1*], p. 355). In the theory of Markov chains the treatment of the general chain is readily deduced from the theory of the chain with only one ergodic class. If assumption (3.5.1) does not hold an analogous treatment will enable one to reduce the problem to the case where (3.5.1) holds. The details are left to the reader who will find it easy to supply them after the theory of this section.

Hitherto we have intentionally said nothing about how the initial state of the channel (at the beginning of a word of length n) is to be determined. Since essentially we are interested in behavior for large n and hence $n \gg a$, it follows from (3.5.1) that it does not matter how the initial state is determined. For, by "wasting" a few initial letters one can always bring the channel to any desired state. Consequently we assume, for simplicity, that the initial state is always chosen at pleasure by the sender and is known to the receiver.

Call any sequence

$$c = (c_1, \ldots, c_n),$$

where c_i, $i = 1, \ldots, n$, is one of $1, \ldots, t$, a "channel sequence" or a "channel n-sequence". Let $u_0 = (x_1, \ldots, x_n)$ be an n-sequence. When we say that the word (n-sequence) u_0 is transmitted according to c we mean that the initial state is c_1, that

$$\phi(c_i, x_i) = c_{i+1}, \quad i = 1, \ldots, n-1 \tag{3.5.2}$$

and that the probability that

$$v(u_0) = (Y_1(u_0), \ldots, Y_n(u_0))$$

shall be the sequence $(y_1, \ldots, y_n)$ is

$$\prod_{i=1}^{n} w(y_i \mid x_i \mid c_i). \tag{3.5.3}$$

If u_0 and c satisfy (3.5.2) we shall call them "compatible".

A code $(n, N(c), \lambda \mid c)$ for a given channel sequence c is a system

$$\{(u_1, A_1), \ldots, (u_{N(c)}, A_{N(c)})\}, \tag{3.5.4}$$

where $u_1, \ldots, u_{N(c)}$ are n-sequences compatible with c, $A_1, \ldots, A_{N(c)}$ are disjoint sets of n-sequences, and

$$P\{v(u_i) \varepsilon A_i \mid c_1\} \geqq 1 - \lambda, \quad i = 1, \ldots, N(c). \tag{3.5.5}$$

The left member of (3.5.5) is the probability of the relation in braces when c_1 is the initial state of the channel. (The succeeding states are then determined. The channel sequence is c because u_i and c are compatible.)

A code (n, N, λ) for the channel is an ensemble of codes $(n, N(c), \lambda \mid c)$ for every channel sequence c. (For some c we may have no code; then $N(c) = 0$.) The length N is defined as

$$N = \Sigma N(c) \tag{3.5.6}$$

where the sum is over all channel sequences c. The reason for this definition is this: Since both sender and receiver always know the state of the channel, each word of the dictionary of words to be transmitted can be coded into a (compatible) *pair*, whose first element is a channel n-sequence and second element an n-sequence.

Let d_1 and d_2 be any two states of the channel, distinct or not. Let

$$B(d_1, d_2) = \{i \mid \phi(d_1, i) = d_2\}. \tag{3.5.7}$$

Consider the d.m.c. whose input alphabet is the set $B(d_1, d_2)$ and whose c.p.f. is $w(\cdot \mid \cdot \mid d_1)$; let $C'(d_1, d_2)$ be its capacity. Define

$$C(d_1, d_2) = \exp_2\{C'(d_1, d_2)\}.$$

If the set $B(d_1, d_2)$ is empty let $C(d_1, d_2) = 0$. Let L be the $(t \times t)$ matrix

$$\{C(d_1, d_2)\}.$$

Let d_1 and d_2 be any two states of the channel, distinct or not. It follows from (3.5.1) that there exists an integer k and a sequence of states $d^{(1)}, \ldots, d^{(k)}$, such that

$$C(d_1, d^{(1)}), C(d^{(1)}, d^{(2)}), \ldots, C(d^{(k)}, d_2)$$

are all greater than zero (hence $\geqq 1$). Hence, by a theorem of FROBENIUS (GANTMACHER [*1*], p. 47, Th. 2), the matrix L possesses a simple positive characteristic root g which is not less in absolute value than any other characteristic root; the characteristic vector of g, say $z = (z_1, \ldots, z_t)$, has all components positive. Moreover (ibid,. p. 74, Th. 2), we may write

$$L = gZMZ^{-1} \tag{3.5.8}$$

where Z is the diagonal matrix with diagonal elements $z_1, \ldots, z_t$, and M is a stochastic matrix. It follows from (3.5.1) that M (or the Markov chain with transition matrix M) has no transient states and contains only one ergodic class (or only one closed set).

We shall show that

$$C = \log g \tag{3.5.9}$$

is the capacity of the channel of this section. We begin by proving

Theorem 3.5.1. Let λ, $0 < \lambda \leqq 1$, be arbitrary. There exists a constant $K_1 > 0$ such that, for any n, there exists a code (n, N, λ) for the channel of this section, with

$$N > g^{n - K_1\sqrt{n}}. \tag{3.5.10}$$

Proof: Let c be any channel sequence. For any pair (i, j), $i, j = 1, \ldots t$, let $R(i, j \mid c)$ be the number of elements k, $k = 1, \ldots, (n-1)$, such that $(c_k, c_{k+1}) = (i, j)$. Fix c and a pair (i, j). We may regard those letters of any n-sequence, sent or received according to c, which have indices k

such that $(c_k, c_{k+1}) = (i, j)$, as constituting words of length $R(i, j \mid c)$, to be transmitted with probability of error $\leqq 1 - (1-\lambda)^{\frac{1}{t^2}}$; the input alphabet is the set $B(i, j)$, and the c.p.f. $w(\cdot \mid \cdot \mid i)$ governs the transmission of each letter. It follows from Theorem 3.2.1 that, for any channel sequence c, there exists a code $(n, N(c), \lambda \mid c)$, with

$$N(c) > \prod_{i,j} [C(i, j)]^{R(i,j \mid c)} \cdot 2^{-t^2 K \sqrt{n}}, \tag{3.5.11}$$

where K is the constant of Theorem 3.2.1. It remains only to evaluate

$$S = \sum_{c} \prod_{i,j} [C(i, j)]^{R(i,j \mid c)}. \tag{3.5.12}$$

Suppose $c = (c_1, \ldots, c_n)$. Then

$$\prod_{i,j} [C(i, j)]^{R(i,j \mid c)} = \prod_{k=1}^{n-1} C(c_k, c_{k+1}).$$

From the definition of matrix multiplication it follows readily that S is the sum of the elements of the matrix L^{n-1}. The sum of the elements of M is t, and so is the sum of the elements of M^{n-1}. Define

$$\alpha = \min_{i,j} \left[\frac{z_i}{z_j}\right]. \tag{3.5.13}$$

It follows from (3.5.8) that

$$S \geqq \alpha\, t g^{n-1}, \tag{3.5.14}$$

from which the theorem readily follows.

We now prove the strong converse:

Theorem 3.5.2. Let λ, $0 \leqq \lambda < 1$, be arbitrary. There exists a constant $K_2 > 0$ such that, for any n, any code (n, N, λ) for the channel of this section satisfies

$$N < g^{n + K_2 \sqrt{n}}. \tag{3.5.15}$$

It is tempting to apply Theorem 3.3.1 directly, in the same way that Theorem 3.2.1 was applied to prove Theorem 3.5.1. This would be incorrect, because the received "words" of length $R(i, j \mid c)$, discussed in the first paragraph of the proof of Theorem 3.5.1, might, because of the coding scheme, give information about transmitted "words" in the same n-sequence other than their own, which would enable codes to be longer. Of course, it turns out that this is not the case, but this has to be proved.

Instead we proceed by applying the *method of proof* of Theorem 3.3.1. This application will be clear as soon as we have described the two points of difference with the proof of Theorem 3.3.1; these are in the definitions of a generated sequence and of a π-sequence.

Let $c = (c_1, \ldots, c_n)$ be any channel sequence, which will be considered fixed in the argument to follow. Let $u_0 = (x_1, \ldots, x_n)$ be any n-sequence. Let $N(i, j, i' \mid u_0)$ be the number of elements k, $k = 1, \ldots, (n-1)$, such that $c_k = i$, $c_{k+1} = j$, and $x_k = i'$; here $i, j = 1, \ldots, t$, and i' is any element of $B(i, j)$. Let

$$\pi = \{\pi(i, j, i')\}, \quad i, j = 1, \ldots, t; \quad i' \varepsilon B(i, j)$$

be any matrix such that $\pi(i, j, i') \geqq 0$, and, for $i, j = 1, \ldots, t$,

$$\sum_{i' \varepsilon B(i,j)} \pi(i, j, i') = 1. \tag{3.5.16}$$

Such matrices will now take the place of π-vectors π. The n-sequence u_0 will be called a π-sequence (always with respect to the fixed c) if

$$\begin{gathered} |N(i, j, i' \mid u_0) - R(i, j \mid c) \cdot \pi(i, j, i')| \\ \leqq 2t\sqrt{aR(i, j \mid c) \cdot \pi(i, j, i')\,(1 - \pi(i, j, i'))}, \end{gathered} \tag{3.5.17}$$

$$i, j = 1, \ldots, t; i' \varepsilon B(i, j).$$

Let $v_0 = (y_1, \ldots, y_n)$ be an n-sequence. Define $N(i, j, i', j' \mid u_0, v_0)$, $i, j = 1, \ldots, t$; $j' = 1, \ldots, a$; $i' \varepsilon B(i, j)$, as the number of elements k, $k = 1, \ldots, (n-1)$, such that

$$c_k = i, \quad c_{k+1} = j, \quad x_k = i', \quad y_k = j'.$$

Let $\delta > 2at$ be a number to be determined in the course of the proof. Call v_0 generated by u_0 if

$$\begin{gathered} |N(i, j, i', j' \mid u_0, v_0) - N(i, j, i' \mid u_0)\, w(j' \mid i' \mid i)| \\ \leqq \delta\,[N(i, j, i' \mid u_0)\, w(j' \mid i' \mid i)\,(1 - w(j' \mid i' \mid i))]^{\frac{1}{2}}, \end{gathered} \tag{3.5.18}$$

$$i, j = 1, \ldots, t; \; i' \varepsilon B(i, j); \; j' = 1, \ldots, a.$$

With these definitions established, it is easy to carry out a proof like that of Theorem 3.3.1. One obtains that

$$N(c) < \prod_{i,j} [C(i, j)]^{R(i,j \mid c)} \cdot 2^{t^2 K' \sqrt{n}}, \tag{3.5.19}$$

which corresponds to (3.5.11). To complete the proof one needs an upper bound on the S of (3.5.12). This is obtained in the same manner as the lower bound (3.5.14). The details are straightforward and are left to the reader.

3.6. The finite-state channel with state calculable only by the sender. We now consider the channel of the previous section, but with the proviso that, at each letter, the state is known to the sender but not to the receiver. (As in Section 3.5, $c_{i+1} = \phi(c_i, x_i)$.) For definiteness we assume that the initial state can be determined by the sender at pleasure.

(If this is not so a trivial change in the argument to follow will suffice to make the theory hold.) We again postulate that (3.5.1) holds. Hence the factorization (3.5.8) holds, with M as described there. To avoid an unessential detail we assume that M is aperiodic. (The reader will easily see how the theory goes through if M is periodic.) The one consequence of aperiodicity which we shall use is that there exists an integer, say d_0, such that the d_0^{th} power of M (and hence of L) has all its elements positive (Doob [1], p. 182, cases f and g; Feller [1], p. 378, Ex. 22). Let d be the smallest such integer with this property. Then, no matter what the initial state of the channel, by sending exactly d suitably chosen letters we can bring the channel into any other prescribed state.

By a "channel n-pair" $g_0 = (u_0, i_0)$, say (the term will be used only in this section), we shall mean a pair consisting of an n-sequence u_0 and a state i_0. To transmit the channel pair g_0 means to put the channel into the initial state i_0 and then to send u_0. (Of course, the succeeding states of the channel are then uniquely determined by ϕ.) As usual, let $v(g_0)$ denote the chance received n-sequence when g_0 is transmitted.

Since the channel sequence c for a word transmitted is not known to the receiver, there can be no question of using codes like the codes $(n, N(c), \lambda \mid c)$ of Section 3.5. Instead we obviously must define a code (n, N, λ) for the channel of the present section as a system

$$\{(g_1, A_1), \ldots, (g_N, A_N)\} \tag{3.6.1}$$

where $g_1, \ldots, g_N$ are channel n-pairs, $A_1, \ldots, A_N$ are disjoint sets of n-sequences, and

$$P\{v(g_i) \varepsilon A_i\} \geqq 1 - \lambda, \quad i = 1, \ldots, N. \tag{3.6.2}$$

Let $C(l)$ be the capacity of the d.m.c. with input alphabet (alphabet of letters to be transmitted) consisting of all channel l-pairs, output alphabet (alphabet of letters received) consisting of the l-sequences, and c.p.f. $w^*(\cdot \mid \cdot)$ defined as follows: Let g' be any channel l-pair, and g'' be any l-sequence. Then

$$w^*(g'' \mid g') = P\{v(g') = g''\}.$$

The quantity $C(l)$ is, at least in principle, computable to within any specified limit of error. We shall show that the capacity of the channel of the present section is

$$C = \inf_l \frac{C(l)}{l}. \tag{3.6.3}$$

The quantity C is not, ex definitione, computable to any specified degree of approximation. However, Theorem 3.6.3 below will, at least in principle, make it so computable.

We shall prove the following theorems:

Theorem 3.6.1. Let $\varepsilon > 0$ and λ, $0 < \lambda \leqq 1$, be arbitrary. For all sufficiently large n there exists a code $(n, 2^{n(C-\varepsilon)}, \lambda)$.

Theorem 3.6.2. Let $\varepsilon > 0$ and λ, $0 \leqq \lambda < 1$, be arbitrary. For all sufficiently large n there does not exist a code $(n, 2^{n(C+\varepsilon)}, \lambda)$.

Theorem 3.6.3. We have, as $l \to \infty$,

$$\lim \frac{C(l)}{l} = C; \tag{3.6.4}$$

for every l we have the upper bound

$$\frac{C(l)}{l} - C \leqq \frac{d \log a}{d + l} \tag{3.6.5}$$

and the bound on the relative error

$$\frac{lC}{C(l)} \geqq 1 - \frac{d}{d + l}. \tag{3.6.6}$$

(Thus C is zero if and only if $C(l)$ is zero for every positive integer l.)

Proof of Theorem 3.6.2. Let l be such that

$$\frac{C(l)}{l} < C + \frac{\varepsilon}{4}. \tag{3.6.7}$$

First suppose $n = kl$, with k an integer. Consider the d.m.c. of which $C(l)$ is the capacity. It follows from Theorem 3.3.1 that, for $0 \leqq \lambda < 1$ and all k sufficiently large, a code (k, N, λ) for this channel satisfies

$$N < 2^{k\left[C(l) + \frac{\varepsilon}{4}\right]} < 2^{n\left[\frac{C(l)}{l} + \frac{\varepsilon}{4}\right]} < 2^{n\left(C + \frac{\varepsilon}{2}\right)}. \tag{3.6.8}$$

Now a code (n, N, λ) for the channel of the present section is surely a code (k, N, λ) for the d.m.c. (but not necessarily vice versa). Hence the theorem follows for n of the form kl.

Suppose now $n = kl + k'$, $1 \leqq k' < l$. Let $n' = (k + 1)\, l$. It follows from (3.6.8) that

$$N < 2^{n'\left(C + \frac{\varepsilon}{2}\right)} < 2^{n(C+\varepsilon)} \tag{3.6.9}$$

for k sufficiently large. This completes the proof of Theorem 3.6.2.

The proof of Theorem 3.6.2. was trivial because of the way we defined C. The proof of Theorem 3.6.1 will also be very easy.

Proof of Theorem 3.6.1. Let l be a positive integer to be chosen later, and suppose $n = kl$, $0 < \lambda \leqq 1$. If k is sufficiently large it follows from Theorem 3.2.1 that there exists a code (k, N, λ) for the d.m.c. whose input alphabet consists of all channel l-pairs and whose capacity is $C(l)$, such that

$$N > 2^{k\left(C(l) - \frac{\varepsilon}{4}\right)}.$$

Let (γ, A) be an element of this code. Then A is a set of n-sequences and $\gamma = (\gamma_1, \ldots, \gamma_k)$ is a sequence of channel l-pairs, $\gamma_j = (u_j, i_j)$, say, $j = 1, \ldots, k$. What keeps this code from being a code for the channel of the present section is that the state of the channel after the transmission of u_j is completed is not necessarily i_{j+1} $(j = 1, \ldots, (k-1))$. Suppose, however, that between u_j and u_{j+1} we transmit d suitably chosen letters at the end of which the channel will be in state i_{j+1}, and "ignore" the received letters which correspond to the inserted d letters. (What "ignore" means is obvious.) We do this for $j = 1, \ldots, (k-1)$, and for every element in the code. The result is a code

$$\left(k(d + l) - d, 2^{k\left(C(l) - \frac{\varepsilon}{4}\right)}, \lambda\right) \tag{3.6.10}$$

for the channel of the present section.

The previous argument is valid for any l. Now choose l so large that

$$\frac{dC}{d + l} < \frac{\varepsilon}{4}. \tag{3.6.11}$$

Write $n' = k(d + l) - d$. It follows from (3.6.3), (3.6.10), and (3.6.11) that, for all sufficiently large n' of the form $k(d + l) - d$, there exists a code (n', N, λ) for the channel of the present section, with

$$N > 2^{n'\left(C - \frac{\varepsilon}{2}\right)}.$$

The proof of the theorem for all other large n' is completed as at the end of the proof of Theorem 3.6.2. This completes the proof of Theorem 3.6.1.

Proof of Theorem 3.6.3. The quantity ε in Theorems 3.6.1. and 3.6.2 was arbitrary. Hence, from (3.6.3), (3.6.10), and Theorem 3.6.2 we obtain, for any l,

$$\frac{C(l)}{d + l} \leqq C \leqq \frac{C(l)}{l}, \tag{3.6.12}$$

from which (3.6.4) and (3.6.6) follow. From (2.2.4) and (3.1.11) we have that $C(l) \leqq l \log a$. Hence (3.6.5) follows. Thus Theorem 3.6.3 is proved.

Remarks

1) Theorem 3.2.1 was brilliantly conjectured by SHANNON [*1*] in 1948. The first proof was given by FEINSTEIN [*1*] in 1954. Essentially different proofs were given in 1957 by SHANNON ([*2*]; see Theorems 7.3.1 and 7.3.2 below) and the author (WOLFOWITZ [*1*]). The proof given here is that of the author.

SHANNON [*1*] also conjectured the weak converse to Theorem 3.2.1 (see Section 7.4 below). The strong converse (Theorem 3.3.1) is due to the author (WOLFOWITZ [*1*]).

Theorem 3.4.1 is due to WEISS [1]. See also ELIAS [1], [2].

2) The introduction of the channels of Section 3.5 and 3.6 is due to SHANNON [2]. The latter proved Theorems 3.5.1 and 3.6.1, and weak converses to these.

Theorems 3.5.2, 3.6.2, and 3.6.3 are new.

3) The "binary erasure" channel has been studied by ELIAS [1], [2]. For this d.m.c. the input alphabet has two symbols, the output alphabet has three symbols, and the c.p.f. is given as follows:

$$w(1 \mid 1) = q, \quad w(2 \mid 1) = 0, \quad w(3 \mid 1) = 1-q,$$
$$w(1 \mid 2) = 0, \quad w(2 \mid 2) = q, \quad w(3 \mid 2) = 1-q.$$

The capacity of this channel is q.

4) The basic steps of the proof of Theorem 3.2.1 are as follows:

a) The idea that it is sufficient to consider only $\bar{\pi}$-sequences and sequences generated by them.

b) The upper bound of Lemma 2.1.6.

c) The upper bound of Lemma 2.1.3.

d) Inequality (3.2.5). This pretty idea is due to FEINSTEIN [1].

From the above it follows that, for $(A_1 \cup \cdots \cup A_N)$ to contain the requisite number of (generated) sequences, N must be sufficiently large.

5) The basic steps of the proof of Theorem 3.3.1 are as follows:

a) The decomposition of the code into subcodes according to π. This enables one to limit consideration to codes all of whose message sequences are π-sequences.

b) The idea that one can limit consideration to generated sequences. (Proof of Lemma 3.3.2.)

c) The upper bound of Lemma 2.1.5.

d) The upper bound of Lemma 2.1.4.

From the above it follows that there are not enough generated sequences for the code to be any longer than the length given in the theorem.

6) The channels studied in Sections 3.5 and 3.6 are essentially channels with memory. As such they could logically be discussed in a later chapter. However, their treatment serves as a useful illustration of the methods of Chapter 3 and it seems pedagogically desirable to include them here.

Any classification of channels, including whatever classification is implied by the chapters of this monograph, is to a considerable extent at least, subjective and arbitrary, and can only be a matter of expediency and convenience. The situation is comparable to the different kinds of dependence which may exist among chance variables.

7) One could also consider the finite-state channel with state unknown to the sender but known to the receiver. Suppose the state is known to the receiver by means of the function ϕ' described in Section 3.5. One could then reasonably describe such a channel as a finite-state channel with state calculable only by the receiver. Whether or not the state of the channel can be determined by a function ϕ is immaterial; the fact that the sender does not know the state of the channel has the consequence that his choice of the n-sequence to transmit can depend only on the word to be transmitted and not on the channel sequence. When ϕ' is a function only of its second argument, the finite-state channel with state calculable only by the receiver is treated in Sections 6.5 and 6.6 below. When ϕ' depends upon both arguments the treatment of the channel is essentially indicated in the paragraph which follows (6.5.1) below.

8) SHANNON [2] gives the capacity of the channel of Section 3.6 as follows: Let s_0 be an arbitrary but fixed state. Let $D(l)$ be the capacity of the d.m.c. such that: a) its input alphabet consists of all channel l-pairs whose second member is s_0, and which are such that, after transmission of any of them over the channel is finished, the channel is back in state s_0; b) its c.p.f. is w^*. Then he proves that the capacity of the channel is

$$\sup_l \frac{D(l)}{l}.$$

Of course this quantity must be the same as C defined in (3.6.3).

9) The treatment of the channels of Sections 3.5 and 3.6 can easily be extended to the case where the state of the channel depends upon several preceding states, not just the last one as in Sections 3.5 and 3.6. In the theory of Markov chains there is a simple standard device for reducing chains of order $m > 1$ to simple chains (of order one); one uses new states which are m-vectors in the old states. The same device can also be used here.

10) In Chapters 2 and 3 we have assumed that both input and output alphabets have the same number of elements. The case where the input alphabet has more elements than the output alphabet is subsumed under the case of equal numbers. The case where the output alphabet has more elements than the input alphabet can be subsumed under the case of equal numbers by adding fictitious elements (which can never be employed) to the input alphabet. Thus the lemmas of Section 2.1 and Theorems 3.2.1 and 3.3.1 hold even when the input and output alphabets do not have the same number of elements. We have already used this fact in Sections 3.5 and 3.6, and will use it again in the chapters to follow. Of course, one could improve the constants in Section 2.1 (and Theorems 3.2.1 and 3.3.1) by taking the actual numbers of input and

output letters into account. Since, however, we have not sought the best constants but only adequate constants, our practice of taking the two alphabets to have the same number of elements has the merit of simplicity.

11) Jacobs [1] has introduced the notion of "almost periodic" channels and proved various results.

4. Compound Channels

4.1. Introduction. Let S be a collection, finite or infinite, of c.p.f.'s $w(\cdot \mid \cdot \mid s)$, where $s \varepsilon S$ is an index which distinguishes different w. (We purposely use the same letter for both the collection of c.p.f.'s and the set of their indices. This is not strictly logical, but it will cause no confusion and will keep down the amount of notation to be introduced. We shall throughout adhere to this practice.) Let $u_0 = (x_1, \ldots, x_n)$ be any word sent and $v(u_0) = (Y_1(u_0), \ldots, Y_n(u_0))$ be the chance word received. Let $v_0 = (y_1, \ldots, y_n)$ be any n-sequence. Define, for every $s \varepsilon S$,

$$P_s\{v(u_0) = v_0\} = \prod_{i=1}^{n} w(y_i \mid x_i \mid s). \tag{4.1.1}$$

A code for S is a set

$$\{(u_1, A_1), \ldots, (u_N, A_N)\}, \tag{4.1.2}$$

where the u_i are n-sequences and the A_i are disjoint sets of n-sequences.

Let

$$P_s\{v(u_i) \varepsilon A_i\} = 1 - \lambda_s(u_i), \quad i = 1, \ldots, N; \quad s \varepsilon S.$$

If

$$\lambda_s(u_i) \leqq \lambda, \quad i = 1, \ldots, N; \quad s \varepsilon S$$

then the code (4.1.2) is called a code (n, N, λ) for the compound discrete memoryless channel S; we will call it for short a code (n, N, λ) for S. In the practical application of such a code the c.p.f. may change *arbitrarily from word to word* (of n letters) in a manner unknown to both sender and receiver, and yet the probability of incorrectly decoding the word sent is $\leqq \lambda$.

We will show that the capacity of such a channel is

$$\overline{C} = \max_{\pi} \inf_{s \varepsilon S} \left[H(\pi' \mid s) - \sum_i \pi_i H(w(\cdot \mid i \mid s)) \right] \tag{4.1.3}$$

where $H(\pi' \mid s)$ and $H(w(\cdot \mid i \mid s))$ are the usual such quantities computed for $w(\cdot \mid \cdot \mid s)$. The quantity in square brackets is continuous in π, uniformly in $w(\cdot \mid \cdot \mid s)$, which justifies our writing max instead of sup. Let $\bar{\pi}$ be a π for which the maximum is attained. π π

Before proceeding to show that $\overline{C}$ is the capacity of the channel S (by proving a coding theorem and a strong converse) we would like to show that $\overline{C}$ is unequal to zero sufficiently often to make the theory interesting. If $\overline{C}$ were zero for most, or even very many, cases of interest, the theory could hardly be interesting. Define $\overline{\overline{C}}$ as

$$\overline{\overline{C}} = \inf_{s \varepsilon S} \max_{\pi} \left[H(\pi' \mid s) - \Sigma\, \pi_i H(w(\cdot \mid i \mid s))\right]. \tag{4.1.4}$$

Thus $\overline{\overline{C}}$ is the infimum of the capacities of the d.m.c.'s whose (single) c.p.f.'s are the various c.p.f.'s of S. Clearly, the capacity of the channel S cannot exceed $\overline{\overline{C}}$, because the capacity of the channel where the c.p.f. varies arbitrarily in S cannot be greater than the capacity of the channel with a single c.p.f., even if that c.p.f. is the "worst" of those in S. We will now show that $\overline{C} = 0$ if and only if $\overline{\overline{C}} = 0$, so that the theory to be developed below is meaningful and not empty.

To show that $\overline{\overline{C}} \geqq \overline{C}$ is trivial. We have, for every $s \varepsilon S$,

$$\left[H(\pi' \mid s) - \sum_i \pi_i H(w(\cdot \mid i \mid s))\right]$$
$$\geqq \inf_s \left[H(\pi' \mid s) - \sum_i \pi_i H(w(\cdot \mid i \mid s))\right].$$

Hence

$$\max_\pi \left[H(\pi' \mid s) - \sum_i \pi_i H(w(\cdot \mid i \mid s))\right] \geqq \overline{C}$$

and

$$\overline{\overline{C}} \geqq \overline{C}. \tag{4.1.5}$$

The inequality sign can obviously hold when the maximizing π are different for different $w(\cdot \mid \cdot \mid s)$.

We now show that $\overline{C} = 0$ implies $\overline{\overline{C}} = 0$. $\overline{C} = 0$ implies that

$$\inf_s \left[H(\pi' \mid s) - \sum_i \pi_i H(w(\cdot \mid i \mid s))\right] = 0 \tag{4.1.6}$$

for every π, and in particular for a π all of whose components are positive. This in turn implies that

$$\inf_s \sum_j \sum_{i,i'} |w(j \mid i \mid s) - w(j \mid i' \mid s)| = 0. \tag{4.1.7}$$

This is so because the infimum of

$$\left[H(\pi') - \sum_i \pi_i H(w(\cdot \mid i))\right]$$

with respect to all c.p.f. $w(\cdot \mid \cdot)$ (not necessarily in S) for which

$$\sum_j \sum_{i,i'} |w(j \mid i) - w(j \mid i')| \geqq c > 0$$

is obviously a positive function of c. But (4.1.7) implies $\overline{\overline{C}} = 0$, by (2.2.10).

4.2. The canonical channel. It will be convenient to approximate S by a sufficiently "dense" set S^* of c.p.f.'s. However, the lemmas of the present section are valid for all c.p.f.'s and we might as well state them accordingly. Let any c.p.f. $w(\cdot \mid \cdot \mid s)$ (not necessarily in S) be approximated by a c.p.f. $w(\cdot \mid \cdot \mid s')$ such that $w(j \mid i \mid s')$ is an integral multiple of $2^{-\sqrt{n}}$ for $i = 1, \ldots, a$, $j = 1, \ldots, a-1$, and

$$|w(j \mid i \mid s) - w(j \mid i \mid s')| \leqq a \cdot 2^{-\sqrt{n}}$$

for $i, j = 1, \ldots, a$. Call the set S^* of c.p.f.'s which approximate the c.p.f.'s of S, the canonical channel, and let S^* also denote its index set. The following lemma is very simple.

Lemma 4.2.1. There exists a constant $K_6 > 0$ with the following property: Let $w(\cdot \mid \cdot \mid s)$ be any c.p.f., not necessarily in S, and $w(\cdot \mid \cdot \mid s')$ an approximating c.p.f. in the sense of the preceding paragraph. Then, for any π,

$$|H(\pi' \mid s) - H(\pi' \mid s')| < K_6 \cdot 2^{-\frac{\sqrt{n}}{2}}$$

$$\left|\sum_i \pi_i H(w(\cdot \mid i \mid s)) - \sum_i \pi_i H(w(\cdot \mid i \mid s'))\right| < K_6 \cdot 2^{-\frac{\sqrt{n}}{2}}.$$

Proof: It is enough to prove the result for large n. In the interval $0 \leqq x \leqq 2^{-\sqrt{n}}$, $-x \log x < 2^{-\frac{\sqrt{n}}{2}}$. In the interval $2^{-\sqrt{n}} \leqq x \leqq 1$, $a \cdot 2^{-\sqrt{n}}$ times the maximum value of the derivative of $-x \log x$ is also less than $2^{-\frac{\sqrt{n}}{2}}$. This proves the lemma.

We now prove

Lemma 4.2.2. Let b be a positive constant. Let $w(\cdot \mid \cdot \mid s)$ be any c.p.f., not necessarily in S, and let $w(\cdot \mid \cdot \mid s')$ be an approximating c.p.f. Let u_0 be any n-sequence and A a set of n-sequences such that

$$P_s\{v(u_0) \varepsilon A\} > b. \tag{4.2.1}$$

Then

$$\left|\frac{P_{s'}\{v(u_0) \varepsilon A\}}{P_s\{v(u_0) \varepsilon A\}} - 1\right| < a_n \tag{4.2.2}$$

where $a_n \to 0$ as $n \to \infty$, and a_n depends only on b and n and not on u_0, A, or $w(\cdot \mid \cdot \mid s)$.

Proof: Suppose first that

$$\min_{i,j} w(j \mid i \mid s) \geqq n^{-2}. \tag{4.2.3}$$

Then the probability, when any given sequence u_0 is sent, that any specified sequence v_0 will be received, is not less than n^{-2n} under $w(\cdot|\cdot|s)$, and not less than $(n^{-2} - a \cdot 2^{-\sqrt{n}})^n$ under $w(\cdot \mid \cdot \mid s')$. We have

$$\frac{\left(n^{-2} - a \cdot 2^{-\sqrt{n}}\right)^n}{n^{-2n}} \to 1 \tag{4.2.4}$$

as $n \to \infty$. If

$$P_s\{v(u_0) = v_0\} > n^{-2n}$$

the approach of

$$\frac{P_{s'}\{v(u_0) = v_0\}}{P_s\{v(u_0) = v_0\}}$$

to 1 is even faster than that of the left member of (4.2.4). Hence the lemma is proved when (4.2.3) holds.

If $w(j_0 \mid i_0 \mid s) \leqq n^{-2}$ then, for any u_0,

$$P_s\{N(i_0, j_0 \mid u_0, v(u_0)) \geqq 1\} \leqq \frac{1}{n}$$

and, for large n,

$$P_{s'}\{N(i_0, j_0 \mid u_0, v(u_0)) \geqq 1\} \leqq \frac{2}{n}.$$

Hence the total contribution, of all received sequences v_0 for which $N(i_0, j_0 \mid u_0, v_0) \geqq 1$ for any pair (i_0, j_0) for which (4.2.3) does not hold, to either $P_s\{v(u_0) \varepsilon A\}$ or $P_{s'}\{v(u_0) \varepsilon A\}$, is not greater than $\frac{2a^2}{n}$, and the influence of this contribution on the ratio in (4.2.2) approaches zero. This proves the lemma.

4.3. A coding theorem. In this section we prove

Theorem 4.3.1. Let λ, $0 < \lambda \leqq 1$, be arbitrary. There exists a positive constant K_0 such that, for any n, there exists a code (n, N, λ) for S with

$$N > \exp_2\left\{n\overline{C} - K_0 \sqrt{n}\right\}. \tag{4.3.1}$$

The number K_0 will in general depend upon λ and a, but it can be chosen independently of S.

Proof: It is clearly sufficient to prove the theorem for n sufficiently large. Let $\lambda' < \lambda$ be any positive number. It follows from Lemma 4.2.2 that, for n sufficiently large, a code with probability of error $\leqq \lambda'$ for the compound discrete memoryless channel S^* is a code with probability of error $\leqq \lambda$ for the compound discrete memoryless channel S.

If an n-sequence v_0 is generated by an n-sequence u_0 with respect to $w(\cdot \mid \cdot \mid s)$, we shall say that v_0 is generated (s) by u_0.

Let $\delta > 2a$ be sufficiently large so that Lemma 2.1.2 holds with $\varepsilon' \leqq \frac{\lambda'}{2}$. Let (4.1.2) be a code (n, N, λ') for the compound channel S^* such that the following conditions are satisfied:

u_i, $i = 1, \ldots, N$, is a $\bar{\pi}$-sequence. (4.3.2)

A_i, $i = 1, \ldots, N$, consists of the union, over S^*, of all n-sequences generated (s) by u_i and not in $A_1 \cup \cdots \cup A_{i-1}$. (4.3.3)

The code is maximal in the sense that it is impossible to add an element (u_{N+1}, A_{N+1}) such that (u_{N+1}, A_{N+1}) satisfies (4.3.2) and (4.3.3) for $i = N + 1$, and $\{(u_1, A_1), \ldots, (u_{N+1}, A_{N+1})\}$ is a code $(n, N + 1, \lambda')$ for the compound channel S^*. (4.3.4)

From (4.3.4) we obtain the following conclusion, exactly as in (3.2.5): Let u_0 be any $\bar{\pi}$-sequence. For some s_0 in S^* we have

$$P_{s_0}\{v(u_0) \text{ is generated } (s_0) \text{ by } u_0 \text{ and belongs to } A_1 \cup \cdots \cup A_N\} > \frac{\lambda'}{2}. \tag{4.3.5}$$

Of course s_0 is a function of u_0. Since there are fewer than $2^{a^2\sqrt{n}}$ indices in S^* it follows from Lemma 2.1.1 that there exists an index s_{00} in S^* and a set B of $\bar{\pi}$-sequences such that

$$P\{X \varepsilon B\} > \frac{3}{4} \cdot 2^{-a^2\sqrt{n}} \tag{4.3.6}$$

and such that for any sequence u_0 in B

$$P_{s_{00}}\{v(u_0) \text{ is generated } (s_{00}) \text{ by } u_0 \text{ and belongs to } A_1 \cup \cdots \cup A_N\} > \frac{\lambda'}{2}. \tag{4.3.7}$$

Hence, from (4.3.6) and (4.3.7), we have

$$P_{s_{00}}\{Y \text{ is generated } (s_{00}) \text{ by some } \bar{\pi}\text{-sequence and belongs to } A_1 \cup \cdots \cup A_N\} > \frac{3\lambda'}{8} \cdot 2^{-a^2\sqrt{n}}. \tag{4.3.8}$$

It follows from (4.3.8) and Lemma 2.1.3 that $A_1 \cup \cdots \cup A_N$ contains at least

$$\frac{3\lambda'}{8} \cdot 2^{-a^2\sqrt{n}} \cdot \exp_2\left\{nH(\bar{\pi}' \mid s_{00}) - K_1 a^3 (1 + \delta) \sqrt{n}\right\} \tag{4.3.9}$$

sequences generated (s_{00}) by some $\bar{\pi}$-sequence, where $K_1 > 0$ is the constant of Lemma 2.1.3.

We now obtain an upper bound on the number of sequences in A_i, $i = 1, \ldots, N$, which are generated (s_{00}) by some $\bar{\pi}$-sequence. The complication which arises here and which is not present in the proof of Theorem 3.2.1 is this: A sequence in A_i which is generated (s) by u_i (some $s \varepsilon S^*$, $s \neq s_{00}$) might also be generated (s_{00}) by some $\bar{\pi}$-sequence, not necessarily u_i. Call an index $s \varepsilon S^*$ for which this can happen "associated" with s_{00}; this terminology will be used only in the present proof in order to avoid repetition. Also let s_{00} be associated with itself.

We now characterize indices s associated with s_{00}. If v_0 is a sequence generated (s_{00}) by a $\bar{\pi}$-sequence then, by (2.1.16),

$$n \sum_j \bar{\pi}_j\, w(i \mid j \mid s_{00}) - 2a^2 \sqrt{n}\, (1+\delta) \sqrt[4]{\sum_j \bar{\pi}_j\, w(i \mid j \mid s_{00})} < N(i \mid v_0)$$

$$< n \sum_j \bar{\pi}_j\, w(i \mid j \mid s_{00}) + 2a^2 \sqrt{n}\, (1+\delta) \sqrt[4]{\sum_j \bar{\pi}_j\, w(i \mid j \mid s_{00})} \tag{4.3.10}$$

for $i = 1, \ldots, a$. If v_0 is also generated (s) by the $\bar{\pi}$-sequence u_i then also

$$n \sum_j \bar{\pi}_j\, w(i \mid j \mid s) - 2a^2 \sqrt{n}\, (1+\delta) \sqrt[4]{\sum_j \bar{\pi}_j w(i \mid j \mid s)} < N(i \mid v_0)$$

$$< n \sum_j \bar{\pi}_j\, w(i \mid j \mid s) + 2a^2 \sqrt{n}\, (1+\delta) \sqrt[4]{\sum_j \bar{\pi}_j\, w(i \mid j \mid s)} \tag{4.3.11}$$

for $i = 1, \ldots, a$. From (4.3.10) and (4.3.11) we obtain, for $i = 1, \ldots, a$, that

$$\left| \sum_j \bar{\pi}_j\, w(i \mid j \mid s_{00}) - \sum_j \bar{\pi}_j\, w(i \mid j \mid s) \right|$$

$$< \frac{2a^2(1+\delta)}{\sqrt{n}} \left(\sqrt[4]{\sum_j \bar{\pi}_j\, w(i \mid j \mid s_{00})} + \sqrt[4]{\sum_j \bar{\pi}_j w\, (i \mid j \mid s)} \right). \tag{4.3.12}$$

Suppose $1 \geqq c > b \geqq 0$, and

$$c - b < \frac{2a^2(1+\delta)}{\sqrt{n}} \left(\sqrt[4]{c} + \sqrt[4]{b} \right). \tag{4.3.13}$$

We wish to bound $|c \log c - b \log b|$. Obviously there is difficulty only for small c. Take $c \leqq \frac{1}{e}$. Then $c \log c < b \log b$. We have, from (4.3.13),

$$b > c - \frac{4a^2(1+\delta)\sqrt[4]{c}}{\sqrt{n}} = q \text{ (say)}. \tag{4.3.14}$$

Suppose first that $q \geqq 0$. Then

$$\begin{aligned} -c \log c + b \log b &< -c \log c + q \log q = -c \log c \\ &+ \left(c - \frac{4a^2(1+\delta)\sqrt[4]{c}}{\sqrt{n}} \right) \log c \\ &+ \left(c - \frac{4a^2(1+\delta)\sqrt[4]{c}}{\sqrt{n}} \right) \log \left(1 - 4a^2(1+\delta)\, c^{-\frac{3}{4}}\, n^{-\frac{1}{2}} \right). \end{aligned} \tag{4.3.15}$$

Now the third term of the right member of (4.3.15) is negative, and we have

$$-c\log c + b\log b < \frac{-4a^2(1+\delta)\sqrt[4]{c}\log c}{\sqrt{n}} \tag{4.3.16}$$

which is less than a positive constant times $a^2(1+\delta)n^{-\frac{1}{2}}$. Now suppose that $q < 0$. Then

$$c < \frac{[4a^2(1+\delta)]^{\frac{4}{3}}}{n^{\frac{2}{3}}}$$

and therefore, for n large,

$$\begin{aligned} &-c\log c + b\log b < -c\log c \\ &< \frac{-[4a^2(1+\delta)]^{\frac{4}{3}}}{n^{\frac{2}{3}}}\log\left\{\frac{[4a^2(1+\delta)]^{\frac{4}{3}}}{n^{\frac{2}{3}}}\right\} \\ &= -\frac{1}{n^{\frac{2}{3}}}[4a^2(1+\delta)]^{\frac{4}{3}}\log[4a^2(1+\delta)]^{\frac{4}{3}} + \frac{2[4a^2(1+\delta)]^{\frac{4}{3}}}{3}\cdot\frac{\log n}{n^{\frac{2}{3}}} \end{aligned} \tag{4.3.17}$$

which is bounded above by an absolute constant times $a^3(1+\delta)^2 n^{-\frac{1}{2}}$.

Now we return to (4.3.12). It and the previous paragraph imply that

$$|H(\bar{\pi}' \mid s_{00}) - H(\bar{\pi}' \mid s)| < \frac{K_7 a^4(1+\delta)^2}{\sqrt{n}}, \tag{4.3.18}$$

where K_7 is a positive constant. Thus (4.3.18) is satisfied by all $s \varepsilon S^*$ associated with s_{00}.

Let $s_{00}^* \varepsilon S^*$ be that index associated with s_{00} for which

$$\sum_i \bar{\pi}_i H(w(\cdot \mid i \mid s_{00}^*)) = \max_s \sum_i \bar{\pi}_i H(w(\cdot \mid i \mid s)), \tag{4.3.19}$$

where the maximum is taken over all indices s associated with s_{00}. From Lemma 2.1.6 we conclude that A_i, $i = 1, \ldots, N$, contains fewer than

$$2^{a^2\sqrt{n}}\cdot\exp_2\left\{n\sum_i \bar{\pi}_i H(w(\cdot \mid i \mid s_{00}^*)) + \sqrt{n}\,(2a+\delta)a^2 K_4\right\} \tag{4.3.20}$$

n-sequences generated (s_{00}) by some $\bar{\pi}$-sequence. $(A_1 \cup \cdots \cup A_N)$ contains fewer than N times this many sequences which are generated (s_{00}) by some $\bar{\pi}$-sequence. Hence, from (4.3.9),

$$\begin{aligned} N &> \frac{3\lambda'}{8}\cdot 2^{-2a^2\sqrt{n}}\exp_2\Big\{n[H(\bar{\pi}' \mid s_{00}) - \sum_i \bar{\pi}_i H(w(\cdot \mid i \mid s_{00}^*))] \\ &\quad -\sqrt{n}\,[K_1 a^3(1+\delta) + K_4(2a+\delta)a^2]\Big\} \\ &> \frac{3\lambda'}{8}\cdot\exp_2\Big\{n[H(\bar{\pi}' \mid s_{00}^*) - \sum_i \bar{\pi}_i H(w(\cdot \mid i \mid s_{00}^*))] \\ &\quad -\sqrt{n}\,[2a^2 + K_1 a^3(1+\delta) + K_4 a^2(2a+\delta) + K_7 a^4(1+\delta)^2]\Big\} \\ &\geqq \frac{3\lambda'}{8}\cdot\exp_2\Big\{n\,\bar{C} - \sqrt{n}\,[2a^2 + K_1 a^3(1+\delta) + K_4(2a+\delta)a^2 \\ &\quad + K_7 a^4(1+\delta)^2 + 2K_6]\Big\} \end{aligned} \tag{4.3.21}$$

which proves the theorem.

4.4. Strong converse. The strong converse of the coding theorem for S states: Let $\varepsilon > 0$ and λ, $0 \leqq \lambda < 1$, be arbitrary. For n sufficiently large there does not exist a code $(n, 2^{n(\overline{C}+\varepsilon)}, \lambda)$ for S. We shall prove the somewhat stronger

Theorem 4.4.1. Let λ, $0 \leqq \lambda < 1$, be arbitrary. There exists a positive constant K_0' such that, for any n, there does not exist a code $(n, 2^{n\overline{C}+K_0'\sqrt{n}}, \lambda)$ for S. K_0' in general depends upon a and λ but it can be chosen independently of the set S.

Proof: As in the proof of Theorem 3.3.1, we may, and do, assume that $\lambda > 0$. The theorem for $\lambda = 0$ will then be true a fortiori.

A code (n, N, λ) for S can be divided into (not necessarily disjoint) subcodes such that in each subcode all the sequences u_i are π-sequences with the same π, all of whose components are multiples of $\frac{1}{n}$. Hence fewer than $(n+1)^a$ subcodes will suffice. We will say that each subcode "belongs" to its corresponding π.

Let π be any π-vector, and $\bar{s} \varepsilon S$ such that

$$H(\pi' \mid \bar{s}) - \sum_i \pi_i H(w(\cdot \mid i \mid \bar{s})) < \overline{C} + \frac{1}{\sqrt{n}}. \tag{4.4.1}$$

The subcode which belongs to π (if any) is a code for S with probability of error $\leqq \lambda$ and hence is surely a code with probability of error $\leqq \lambda$ (or the d.m.c. with (single) c.p.f. $w(\cdot \mid \cdot \mid \bar{s})$. As the latter it must, by f4.4.1) and Lemma 3.3.2 be shorter than

$$\exp_2\left\{n\overline{C} + \sqrt{n}\,[1 + K_2' a^3 (2a + \delta)]\right\}, \tag{4.4.2}$$

where K_2' is the constant of Lemma 3.3.2, and δ is as in Lemma 3.3.2 for all the c.p.f.'s in S. $\left(\text{It suffices to take } \delta = \max\left[a\sqrt{\frac{2}{1-\lambda}}, 2a\right].\right)$ Since both K_2' and δ do not depend on π we have the length of the entire code is less than

$$(n+1)^a \cdot \exp_2\left\{n\overline{C} + \sqrt{n}\,[1 + K_2' a^3 (2a + \delta)]\right\} \tag{4.4.3}$$

which proves the theorem.

4.5. Compound d.m.c. with c.p.f. known only to the receiver or only to the sender. Suppose we consider the channel of Section 4.1 with this difference, that the receiver, but not the sender, knows the c.p.f. $w(\cdot \mid \cdot \mid s)$ being used for any word; call this channel S_1. The sequence being sent cannot then be affected by this knowledge, but the decoding system can be. A code (n, N, λ) for S_1 is then a set of codes T_s, $s \varepsilon S$. For each s in S, T_s is a code (n, N, λ) for the d.m.c. with (single) c.p.f. $w(\cdot \mid \cdot \mid s)$, thus:

$$T_s = \{(u_1, A_1(s)), \ldots, (u_N, A_N(s))\} \tag{4.5.1}$$

where u_i, $i = 1, \ldots, N$, is the same for all s, but $A_i(s)$ is a function of s. Of course $A_1(s), \ldots, A_N(s)$ are disjoint for any one s, but $A_i(s)$ and $A_j(s')$ $(i \neq j)$ need not be disjoint for $s \neq s'$. If $A_i(s)$ does not depend on s then we have a code (n, N, λ) for S.

The practical use of a code (n, N, λ) for S_1 is this: When the sender wishes to send the ith word he sends u_i. When the receiver knows that $w(\cdot \mid \cdot \mid s)$ is the c.p.f. being used by the system and the received sequence lies in $A_j(s)$ he concludes that the jth word u_j has been sent. The probability that any word will be incorrectly understood is $\leqq \lambda$.

Of course the capacity of this channel must be $\geqq \overline{C}$. We will show that it is actually $\overline{C}$.

Theorem 4.5.1. Theorem 4.3.1 holds verbatim if we replace S by S_1.

Proof: A code (n, N, λ) for S is a code (n, N, λ) for S_1.

Theorem 4.5.2. Theorem 4.4.1 holds verbatim if we replace S by S_1.

Proof: Proceed exactly as in the proof of Theorem 4.4.1. Let π be any π-vector. Consider the subcode of the code (n, N, λ) for S_1 which belongs to π, and select from this subcode the code $T_{\bar{s}}$ for $\bar{s}$ which satisfies (4.4.1). Then the length of $T_{\bar{s}}$ and hence the length of the subcode itself satisfy (4.4.2). Hence, exactly as in Theorem 4.4.1, (4.4.3) holds, and the theorem is proved.

Suppose we now consider the channel S with this difference, that the sender but not the receiver knows the c.p.f. $w(\cdot \mid \cdot \mid s)$ being used for any word; call this channel S_2. In this channel the decoding system cannot be changed but the sequence sent can be made to depend on s. A code (n, N, λ) for S_2 is then a set of codes T'_s, $s \varepsilon S$; for each s in S, T'_s is a code (n, N, λ) for the d.m.c. with (single) c.p.f. $w(\cdot \mid \cdot \mid s)$, thus:

$$T'_s = \{(u_1(s), A_1), \ldots, (u_N(s), A_N)\}. \tag{4.5.2}$$

Here $u_i(s)$ depends upon s but A_i is the same for all s. Of course $A_1, \ldots, A_N$ are disjoint.

The practical use of this system is obvious. When the sender knows that $w(\cdot \mid \cdot \mid s)$ is the c.p.f. being used by the system and he wishes to send the ith word he sends $u_i(s)$. When the receiver receives a sequence in A_j he concludes that the jth word has been sent.

The capacity of this channel will be shown to be $\overline{\overline{C}}$. Thus $\overline{\overline{C}} - \overline{C}$ is the gain in capacity which comes from the sender's knowing the c.p.f. The following heuristic explanation of why the receiver's knowing the c.p.f. does not cause a corresponding gain in capacity is due to Dr. H. KESTEN (oral communication): Already for channel S the receiver can know the c.p.f. with probability arbitrarily close to one when n is sufficiently large. It is sufficient for sender and receiver to arrange that each u_i sent be prefaced by an agreed upon number of repetitions of the letters $1, 2, \ldots, a$; these together with u_i are to constitute a word transmitted in the new code. From the distribution of the initial letters received the receiver can, with high probability, correctly infer the c.p.f. being used. When n is sufficiently large (and ours is essentially a theory for large n) the letters "wasted" at the beginning of each word constitute a negligible fraction of n which affects the code length only by multiplication by $o(\exp_2 \sqrt{n})$.

Theorem 4.5.3. The coding Theorem 4.3.1 holds if we replace S by S_2 and $\bar{C}$ by $\bar{\bar{C}}$.

Proof: We shall lean heavily on the proof of Theorem 4.3.1. As in the latter, we proceed for the canonical channel S^*, and for $\lambda' < \lambda$.

For each $s \varepsilon S^*$ define $\pi(s)$ to be any π-vector for which

$$\begin{aligned} \bar{\bar{C}} + \frac{2K_6}{\sqrt{n}} &> [H(\pi'(s) \mid s) - \sum_i \pi_i(s) H(w(\cdot \mid i \mid s))] \\ &> \bar{\bar{C}} - \frac{2K_6}{\sqrt{n}}. \end{aligned} \tag{4.5.3}$$

The existence of such a π-vector is guaranteed by Lemma 4.2.1.

We now sketch the proof very briefly. Proceed as in the proof of Theorem 4.3.1. Each $u_i(s)$ is a $\pi(s)$-sequence. For every s in S^*, A_i contains all sequences which are generated (s) by $u_i(s)$ and are not in $A_1 \cup \cdots \cup A_{i-1}$. In each of the generated sequences the proportion of each of the elements $1, \ldots, a$ differs by less than a suitable multiple of $\frac{1}{\sqrt{n}}$ from the corresponding component of $\pi'(s)$. If, for every pair s_1 and s_2 in S^*, we have $\pi'(s_1)$ and $\pi'(s_2)$ essentially different (i.e., one or more components differ by more than a suitable multiple of $\frac{1}{\sqrt{n}}$), then the proof goes forward as in Theorems 3.2.1 and 4.3.1. The prolongation of the code is stopped for some index, and since this index must satisfy the right inequality in (4.5.3), the theorem follows. Suppose that, for some pair s_1 and s_2 in S^*, $\pi'(s_1)$ and $\pi'(s_2)$ differ, in every component, by less than this suitable multiple of $\frac{1}{\sqrt{n}}$. Then it follows from (4.5.3) that

$$\left| \sum_i \pi_i(s_1) H(w(\cdot \mid i \mid s_1)) - \sum_i \pi_i(s_2) H(w(\cdot \mid i \mid s_2)) \right|$$

is less than a suitable multiple of $\frac{1}{\sqrt{n}}$. From this we get the analogue of (4.3.20). The theorem follows.

Theorem 4.5.4. Theorem 4.4.1 holds if we replace S by S_2 and $\bar{C}$ by $\bar{\bar{C}}$.

Proof: Apply Theorem 3.3.1 to the individual codes T'_s; $\bar{\bar{C}}$ is the infimum of the capacities.

4.6. Channels where the c.p.f. for each letter is stochastically determined. Suppose given t c.p.f.'s, $w(\cdot \mid \cdot \mid s)$, $s = 1, \ldots, t$, called "states" of the channel, and a probability t-vector $g = (g(1), \ldots, g(t))$. In this section we will study several channels in each of which the c.p.f. (state of the channel) which governs the transmission of a *letter* is determined by chance, independently of all transmitted letters, of all previously received letters, and of all previous states, with probability $g(s)$ that $w(\cdot \mid \cdot \mid s)$ is the c.p.f., $s = 1, \ldots, t$. The vector g is known to both sender and receiver. We shall study four channels, where:

I. Neither the sender nor the receiver knows the c.p.f. which governs the transmission of any letter.

II. Both sender and receiver know which c.p.f. governs the transmission of each letter. It does not matter when the receiver knows this, since he will not "decode" a received word (n-sequence) until the entire word has been received. As for the sender's knowing the c.p.f., we consider two possibilities: 1) channel IIa: The c.p.f. which governs the transmission of any letter is known to the sender before that letter is sent, but only after all the preceding letters of the word have been sent. 2) channel IIb: The c.p.f. for *every* letter is known to the sender before the transmission of the word (of any letter) begins.

III. The receiver, but not the sender, knows the c.p.f. which governs the transmission of each letter.

IV. The sender, but not the receiver, knows the c.p.f. which governs the transmission of each letter. He knows the c.p.f. which governs the transmission of any letter before that letter is sent, but only after all preceding letters of the word have been sent.

Channel I. This is obviously a d.m.c. with c.p.f. $w(\cdot \mid \cdot)$ given by

$$w(k \mid i) = \sum_{s=1}^{t} g(s)\, w(k \mid i \mid s) \tag{4.6.1}$$

for $i, k = 1, \ldots, a$. All the results for a d.m.c. apply.

Channel II. Let $C(s)$, $s = 1, \ldots, t$, be the capacity of the d.m.c. with the single c.p.f. $w(\cdot \mid \cdot \mid s)$. Define C_2 by

$$C_2 = \sum_{s=1}^{t} g(s)\, C(s). \tag{4.6.2}$$

We then have

Theorem 4.6.1. Let λ, $0 < \lambda < 1$, be any number. There exists a constant $K > 0$ such that, for either channel IIa or channel IIb and for any n there exists a code $(n, 2^{nC_2 - K\sqrt{n}}, \lambda)$, and there does not exist a code $(n, 2^{nC_2 + K\sqrt{n}}, \lambda)$.

The proof of this theorem is so simple that we omit it. We have also omitted the formal statement of what a code for this channel is. Not only is the idea of a code obvious, but its formal description sufficiently resembles the descriptions to be given for channels III and IV that it is unnecessary to give it. The proof of the theorem rests on the fact that, with large probability, the number of letters in any word whose transmission is governed by the c.p.f. $w(\cdot \mid \cdot \mid s)$, $s = 1, \ldots, t$, differs from $ng(s)$ by at most a suitable multiple of $\sqrt{n}$.

Before embarking on the more serious task of studying channels III and IV we need a few preliminary remarks. Define a channel n-sequence as in Section 3.5. Let

$$\gamma = (\gamma_1, \ldots, \gamma_n)$$

be a sequence of n independent chance variables, whose common distribution is the t-vector g. Thus γ is a chance channel n-sequence. We will say that, when γ is any channel n-sequence c, the distribution of the ith letter ($i = 1, \ldots, n$) of the word to be transmitted will be governed by the c.p.f. $w(\cdot \mid \cdot \mid c_i)$. Let $u_0 = (x_1, \ldots, x_n)$ and $v_0 = (y_1, \ldots, y_n)$ be any n-sequences. Let $v(u_0)$ be, as usual, the chance received sequence when u_0 is the sequence transmitted. Thus we have:

$$P\{\gamma = c\} = \prod_{i=1}^{n} g(c_i) \tag{4.6.3}$$

$$P\{v(u_0) = v_0 \mid \gamma = c\} = \prod_{i=1}^{n} w(y_i \mid x_i \mid c_i) \tag{4.6.4}$$

$$P\{(\gamma, v(u_0)) = (c, v_0)\} = \prod_{i=1}^{n} g(c_i)\, w(y_i \mid x_i \mid c_i). \tag{4.6.5}$$

Channel III. Call a pair (c, v_0), where c is a channel n-sequence and v_0 an n-sequence, a bi-sequence. (This term will be used only in the study of Channel III, and only very briefly.) Suppose given a set A of bi-sequences. Let $A(c)$ denote the totality of elements of A which have c as the first member of the pair. Then

$$P\{(\gamma, v(u_0))\, \varepsilon A\} = \sum_{c} P\{(c, v(u_0))\varepsilon A(c) \mid \gamma = c\} \cdot P\{\gamma = c\} \tag{4.6.6}$$

where the summation is over all channel sequences c.

A code (n, N, λ) for channel III is a set

$$\{(u_1, A_1), \ldots, (u_N, A_N)\}, \tag{4.6.7}$$

where $u_1, \ldots, u_N$ are n-sequences, $A_1, \ldots, A_N$ are disjoint sets of bi-sequences, and

$$P\{(\gamma, v(u_i))\varepsilon A_i\} \geqq 1 - \lambda \tag{4.6.8}$$

for $i = 1, \ldots, N$. The practical use of such a code is as follows: When the sender wants to send the ith word he sends u_i. When the receiver knows that a word has been transmitted according to the channel sequence c, that the sequence actually received is v_0, and that (c, v_0) lies in A_j, he concludes that the jth word has been sent. If (c, v_0) does not lie in $A_1 \cup \cdots \cup A_N$ the receiver may draw any conclusion he wishes about the word that has been sent. The probability that any word transmitted will be incorrectly decoded is $\leq \lambda$.

Let G be the set of pairs (j, k), $j = 1, \ldots, t$; $k = 1, \ldots, a$. Clearly the channel behaves like a d.m.c. with input alphabet $1, \ldots, a$, output alphabet G, and c.p.f. $q_1(\cdot, \cdot \mid \cdot)$ defined as

$$q_1(j, k \mid i) = g(j)\, w(k \mid i \mid j), \quad i, k = 1, \ldots, a; \; j = 1, \ldots, t. \tag{4.6.9}$$

Let π be any probability a-vector. Then the distribution on the output alphabet G which is completely analogous to the vector π' of (2.1.8) is the distribution q_2 defined as

$$q_2(j, k \mid \pi) = \sum_i \pi_i q_1(j, k \mid i). \tag{4.6.10}$$

Hence C_3, given by

$$C_3 = \max_\pi \left[H(q_2(\cdot, \cdot \mid \pi)) - \sum_i \pi_i H(q_1(\cdot, \cdot \mid i)) \right] \tag{4.6.11}$$

is the capacity of channel III, and we have

Theorem 4.6.2. Let λ, $0 < \lambda < 1$, be any number. There exists a constant $K > 0$ such that, for all n, there exists a code $(n, 2^{nC_3 - K\sqrt{n}}, \lambda)$ for channel III, and there does not exist a code $(n, 2^{nC_3 + K\sqrt{n}}, \lambda)$ for channel III.

Channel IV. A code for this channel will be defined with the aid of a function d, which we now describe. Write

$$c^{(j)} = (c_1, \ldots, c_j),$$
$$\gamma^{(j)} = (\gamma_1, \ldots, \gamma_j), \quad j = 1, \ldots, n.$$

Thus $\gamma = \gamma^{(n)}$, $c = c^{(n)}$. The value of $d(i, j, c^{(j-1)}, c_j)$ is defined for every $i = 1, \ldots, N$; $j = 1, \ldots, n$, and every channel sequence c (actually d is a function only of the first j components of c), and lies in $\{1, \ldots, a\}$. For fixed i call the function $d(i, \cdot, \cdot, \cdot)$ the ith component of d.

We shall need the following chance variables in the description of a code and in the proof to follow: The chance variable u is to have values in the set $\{1, \ldots, N\}$, and distribution Q' (say); it is to be independent of γ, which has already been defined. The chance variables $V(j)$, $j = 1, \ldots, n$, are, for fixed u and γ, distributed independently of each other with the following distributions:

$$P\{V(j) = k \mid u = i, \gamma = c\} = w(k \mid d(i, j, c^{(j-1)}, c_j) \mid c_j) \tag{4.6.12}$$

for $k = 1, \ldots, a$; $i = 1, \ldots, N$, and any channel sequence c. Finally, define

$$v = (V(1), \ldots, V(n)).$$

A code (n, N, λ) for channel IV is a system

$$\{(d(1,\cdot,\cdot,\cdot), A_1), \ldots, (d(N,\cdot,\cdot,\cdot), A_N)\}, \qquad (4.6.13)$$

where $d(1,\cdot,\cdot,\cdot), \ldots, d(N,\cdot,\cdot,\cdot)$ are the components of a function d, the A_i are disjoint sets of n-sequences, and the following holds for $i = 1, \ldots, N$:

$$P\{v \,\varepsilon\, A_i \mid u = i\} \geqq 1 - \lambda. \qquad (4.6.14)$$

The practical application of the code is as follows: If the sender wants to send the ith word, and c is the channel sequence which governs its transmission, he sends successively the symbols

$$d(i, 1, -, c_1), d(i, 2, c^{(1)}, c_2), \ldots, d(i, n, c^{(n-1)}, c_n). \qquad (4.6.15)$$

(The jth of these symbols, $j = 1, \ldots, n$, depends only on $c^{(j)}$, so that the code restriction implicit in Channel IV is satisfied.) If the receiver receives a sequence in A_s he concludes that the sth word has been sent. If the sequence received is not in $A_1 \cup \cdots \cup A_N$ the receiver may draw any conclusion he wishes about the word which has been sent. The probability that any word transmitted will be incorrectly decoded is $\leqq \lambda$.

Consider a d.m.c. whose output alphabet is $1, \ldots, a$, and whose input alphabet (call it G') consists of all t-sequences whose elements are in the alphabet $1, \ldots, a$ (so that G' has a^t elements). Let the c.p.f. $w^*(\cdot \mid \cdot)$ of this channel be defined as follows: When the input letter which is the t-sequence $(b_1, \ldots, b_t)$ is transmitted, the probability that k is received is given by

$$w^*(k \mid b_1, \ldots, b_t) = \sum_{j=1}^{t} g(j)\, w(k \mid b_j \mid j). \qquad (4.6.16)$$

Let C_4 be the capacity of this d.m.c., to which we shall refer as the d.m.c. associated with channel IV.

In order to give an explicit formula for C_4, order the t-sequences of G' in some arbitrary but fixed manner, thus: $\{(a_1^{(i)}, \ldots, a_t^{(i)}), i = 1, \ldots, a^t\}$. Let $q_3(\cdot \mid i)$ be the distribution on $\{1, \ldots, a\}$ which depends on i, $i = 1, \ldots, a^t$, as a parameter, and is defined thus:

$$q_3(k \mid i) = \sum_{j=1}^{t} g(j)\, w(k \mid a_j^{(i)} \mid j), \quad k = 1, \ldots, a. \qquad (4.6.17)$$

The relation between (4.6.16) and (4.6.17) is obviously this:

$$q_3(k \mid i) = w^*(k \mid a_1^{(i)}, \ldots, a_t^{(i)}). \qquad (4.6.18)$$

We use q_3 simply because it is more convenient and suggestive. The operation of the d.m.c. associated with channel IV may be thought of in the following way: Suppose the ith symbol (t-sequence) of G' is to be sent over the d.m.c. The sender "sends" it over channel IV as follows:

He learns which c.p.f. is to govern the transmission of the ith letter over channel IV. Suppose this is the jth c.p.f. He then sends $a_j^{(i)}$ over channel IV. The probability of receiving the letter k over channel IV is the same as the probability of receiving k over the d.m.c.

Let π be a probability a^t-vector, and define the distribution $q_4(\cdot \mid \pi)$ on the set $\{1, \ldots, a\}$, where q_4 depends on π as a parameter, as follows:

$$q_4(k \mid \pi) = \sum_{i=1}^{a^t} \pi_i q_3(k \mid i). \tag{4.6.19}$$

Obviously the distribution q_4 bears the same relation to q_3 as π' does to π in Chapter 3. We see that

$$C_4 = \max_{\pi} \left[H(q_4(\cdot \mid \pi)) - \sum_{i=1}^{a^t} \pi_i H(q_3(\cdot \mid i))\right]. \tag{4.6.20}$$

In view of the discussion of the above d.m.c. we may leave the proof of the following theorem as a trivial exercise for the reader:

Theorem 4.6.3. Let λ, $0 < \lambda \leqq 1$, be arbitrary. There exists a constant $K' > 0$ such that, for any n, there exists a code $(n, 2^{nC_4 - K'\sqrt{n}}, \lambda)$ for channel IV.

The strong converse to the coding Theorem 4.6.3 holds and its proof is almost identical with that of Theorem 4.9.1 below. We shall therefore omit it and prove a weak converse (Theorem 4.6.4) instead. Although this is a weaker theorem its proof is of interest because its techniques can be applied to many other channels, and it therefore seems worthwhile to give it here. For a discussion of the difference between a weak and a strong converse see Chapter 7 below.

Theorem 4.6.4. For any n a code (n, N, λ) for channel IV satisfies

$$\log N < \frac{nC_4 + 1}{1 - \lambda}. \tag{4.6.21}$$

The proof, to be given in the next section, will be intelligible only to a reader familiar with the contents of Chapter 7. Consequently the reader may at this point prefer to pass to Chapter 7 and return to Section 4.7 later.

4.7. Proof of Theorem 4.6.4. We remind the reader of the connection between channel IV and its associated d.m.c. Of course the weak (even the strong) converse applies to the associated d.m.c. However, this does not prove our theorem, for the reason that, while a code for the d.m.c. is a code for channel IV, a code for channel IV need not be a code for the d.m.c. To see this, note that the transmission over channel IV of a letter of the d.m.c. (element of G') requires that the decision as to which symbol is actually to be sent (over channel IV), be made solely on the basis of what c.p.f. governs the transmission of *this* symbol. On the other hand, this decision could, in the general case of channel IV, be based on this c.p.f. *and the c.p.f.'s of all previously transmitted letters* which (the c.p.f.'s) are known to the sender. Thus, as far as weak converses are concerned, Theorem 4.6.4 says essentially that nothing is to be gained by taking into consideration the c.p.f.'s of the previously transmitted letters.

Let Q'_{0n} be the distribution (stochastic n-input) which assigns probability $\frac{1}{N}$ to each of the messages (numbered from 1 to N) which can be sent over channel IV by means of the code (4.6.13). (The notation is chosen in accord with the notation of Chapter 7.) We will prove that

$$H(Q'_{0n}) - H(Q'_{0n} \mid Q''_{0n}) \leqq n C_4. \tag{4.7.1}$$

From this and Theorem 7.4.1 the desired result follows as in (7.4.7) and (7.4.8).

Let u be the chance variable defined in the discussion of channel IV in Section 4.6, with $Q' = Q'_{0n}$, and Q'_{0n} as just defined in the preceding paragraph. Let v be as defined in connection with the definition of u in Section 4.6. Write

$$v^{(j)} = (V(1), \ldots, V(j)), \quad j = 1, \ldots, n.$$

Of course

$$H(Q'_{0n}) = H(u) = \log N$$
$$H(Q'_{0n} \mid Q''_{0n}) = H(u \mid v).$$

We now show that

$$H(u \mid v^{(j-1)}) - H(u \mid v^{(j)}) \leqq C_4, \quad j = 1, \ldots, n. \tag{4.7.2}$$

(For $j = 1$ this means that

$$H(u) - H(u \mid v^{(1)}) \leqq C_4.)$$

Obviously (4.7.2) implies (4.7.1).

From (4.6.12) we obtain that

$$\begin{aligned} &P\{V(j) = k \mid u = i, \gamma^{(j-1)} = c^{(j-1)}\} \\ &\quad = P\{V(j) = k \mid u = i, \gamma^{(j-1)} = c^{(j-1)}, v^{(j-1)} = v_0^{(j-1)}\} \end{aligned} \tag{4.7.3}$$

for any i, j, k, $c^{(j)}$, and $v_0^{(j-1)}$. Hence

$$\begin{aligned} H(V(j) \mid u, \gamma^{(j-1)}) &= H(V(j) \mid u, \gamma^{(j-1)}, v^{(j-1)}) \\ &\leqq H(V(j) \mid u, v^{(j-1)}) \end{aligned} \tag{4.7.4}$$

by (2.2.15). From (2.2.2) we have

$$\begin{aligned} &H(u \mid v^{(j-1)}) - H(u \mid v^{(j-1)}, V(j)) \\ &\quad = H(V(j) \mid v^{(j-1)}) - H(V(j) \mid u, v^{(j-1)}). \end{aligned} \tag{4.7.5}$$

From (2.2.15) we also obtain

$$\begin{aligned} &H(V(j) \mid v^{(j-1)}) - H(V(j) \mid u, v^{(j-1)}) \\ &\leqq H(V(j)) - H(V(j) \mid u, v^{(j-1)}). \end{aligned} \tag{4.7.6}$$

From (4.7.4) we have

$$\begin{aligned} &H(V(j)) - H(V(j) \mid u, v^{(j-1)}) \\ &\leqq H(V(j)) - H(V(j) \mid u, \gamma^{(j-1)}). \end{aligned} \tag{4.7.7}$$

Hence, from (4.7.5)—(4.7.7) we obtain

$$\begin{aligned} &H(u \mid v^{(j-1)}) - H(u \mid v^{(j)}) \\ &\leqq H(V(j)) - H(V(j) \mid u, \gamma^{(j-1)}) \end{aligned} \tag{4.7.8}$$

for $j = 1, \ldots, n$. It remains only to show that the right member of (4.7.8) is $\leqq C_4$.

Let Z_1 be a chance variable with values in G' and distribution π. Let Z_2 be a chance variable with values in $\{1, \ldots, a\}$ and conditional distribution as follows:

$$P\{Z_2 = j \mid Z_1 = i\} = q_3(j \mid i), \quad i = 1, \ldots, a^t; j = 1, \ldots, a.$$

Obviously, from (4.6.20),

$$H(Z_2) - H(Z_2 \mid Z_1) \leqq C_4. \tag{4.7.9}$$

For any i, j, and $c^{(j-1)}$ let $\alpha(i, j, c^{(j-1)})$ be the index of the following member of G':

$$(d(i, j, c^{(j-1)}, 1), \ldots, d(i, j, c^{(j-1)}, t)). \tag{4.7.10}$$

Then, for any i, j, k, and $c^{(j-1)}$,

$$\begin{aligned} &P\{V(j) = k \mid u = i, \gamma^{(j-1)} = c^{(j-1)}\} \\ &\quad = P\{Z_2 = k \mid Z_1 = \alpha(i, j, c^{(j-1)})\}. \end{aligned} \tag{4.7.11}$$

Hence if, for every i, j, and $c^{(j-1)}$, the vector π assigns to $\alpha(i, j, c^{(j-1)})$ a probability equal to

$$\sum_{i_0, c_0^{(j-1)}} P\{u = i_0, \gamma^{(j-1)} = c_0^{(j-1)}\}$$

where the summation is over all i_0, $c_0^{(j-1)}$ such that

$$\alpha(i_0, j, c_0^{(j-1)}) = \alpha(i, j, c^{(j-1)}),$$

the distribution of Z_2 will be the same as the distribution of $V(j)$, $H(Z_2) = H(V(j))$, and $H(Z_2|Z_1) = H(V(j) \mid u, \gamma^{(j-1)})$. The desired bound on the right member of (4.7.8) then follows from (4.7.9). This proves the theorem.

4.8. The d.m.c. with feedback. Consider the d.m.c. with the notation of Section 3.1 and hence the capacity C. Suppose the channel modified "physically" so that the sender can "see" the (chance) received letters as they are received by the receiver. The sender sees the jth received letter after sending the jth transmitted letter and before sending the $(j+1)^{\text{st}}$ transmitted letter. Of course, the sender can communicate with the receiver only through the channel. Obviously, even if the sender ignores the "feedback", i.e., the information continually being supplied to him of what letters the receiver has already received, Theorem 3.2.1 applies and assures the existence of a code $(n, 2^{nC-K\sqrt{n}}, \lambda)$ for the channel. The question obviously arises whether the feedback could not be used to achieve a longer code. Theorem 4.8.1 below says essentially that, for small λ, the answer to the question is no. This is also the answer for all $\lambda (0 \leqq \lambda < 1)$ (see Theorem 4.9.1).

We have described the channel in physical terms for the sake of brevity. A full formal description seems unnecessary, because the channel sufficiently resembles channel IV of Section 4.6. Whenever possible the notation of Sections 4.6 and 4.7 will be adopted below either as it stands or with suitable modifications.

The channel of the present section is not a compound channel. (See, however, the remarks at the end of the chapter.) It is discussed here only because of its basic similarity to channel IV of Section 4.6 and because the proof of Theorem 4.8.1 below is essentially the same as that of Theorem 4.6.4. (Consequently this proof also presupposes a familiarity with the contents of Chapter 7.)

Suppose that the ith word of the code is to be sent $(i = 1, \ldots, N)$, that the first $(j-1)$ letters of the word have already been sent $(j = 1, \ldots, n)$, and the chance received letters $Z_1, \ldots, Z_{j-1}$ (known to both sender and receiver) have already been received. The next (j^{th}) letter to be sent is, similarly to the situation described in Section 4.6 under channel IV, given by $f(i, j, Z_1, \ldots, Z_{j-1})$, where f is a function whose range is the set $\{1, \ldots, a\}$ and whose domain is the Cartesian product of the sets $\{1, \ldots, N\}$, $\{1, \ldots, n\}$, and the set $\{1, \ldots, a\}$ taken $(j-1)$ times, respectively. The formal definition of a code, similar to that of (4.6.13), in which the sender uses the function f (which is the analogue of d) to take account of the feedback, is easy to give and we leave its formal description to the reader. We now prove

Theorem 4.8.1. Any code (n, N, λ) for the channel of the present section satisfies

$$\log N < \frac{nC+1}{1-\lambda}.$$

Let u be the chance variable of the proof of Theorem 4.6.4 (its distribution is Q'_{0n}). Let Y_j, $j = 1, \ldots, n$, be the chance letter received when the jth letter of u is sent. Write

$$Y^{(j)} = (Y_1, \ldots, Y_j), \quad j = 1, \ldots, n$$

and for brevity write $Y = Y^{(n)}$. The proof will consist of repeating, with appropriate variations, the argument of Theorem 4.6.4. The following table of analogous elements will help make the argument easy to follow.

Theorem 4.6.4	*Theorem 4.8.1*
C_4	C
u	u
v	Y
$v^{(j)}$	$Y^{(j)}$
$V(j)$	Y_j
d	f
Z_1	M_j

As in (4.7.1), it is sufficient to prove

$$H(Q'_{0n}) - H(Q'_{0n} \mid Q''_{0n}) \leqq nC. \tag{4.8.1}$$

Then it follows that, as in (4.7.2), it is sufficient to prove that

$$H(u \mid Y^{(j-1)}) - H(u \mid Y^{(j)}) \leqq C, \quad j = 1, \ldots, n. \tag{4.8.2}$$

From (2.2.2) and (2.2.15) we obtain that

$$\begin{aligned} &H(u \mid Y^{(j-1)}) - H(u \mid Y^{(j)}) \\ &= H(Y_j \mid Y^{(j-1)}) - H(Y_j \mid u, Y^{(j-1)}) \\ &\leqq H(Y_j) - H(Y_j \mid u, Y^{(j-1)}). \end{aligned} \tag{4.8.3}$$

Let M_j, $j = 1, \ldots, n$, be the chance variable $f(u, j, Y^{(j-1)})$. Hence, for $j = 1, \ldots, n$,

$$H(Y_j) - H(Y_j \mid u, Y^{(j-1)}) = H(Y_j) - H(Y_j \mid M_j) \leqq C \tag{4.8.4}$$

by (3.1.11). This proves the theorem.

We have given the proof of Theorem 4.8.1 because it is an interesting parallel to that of Theorem 4.6.4. The following theorem (strong converse) actually holds.

Theorem 4.8.2. Any code (n, N, λ) for the channel of section 4.8 satisfies

$$N < 2^{nC + K\sqrt{n}},$$

where K depends on λ but not on n.

This theorem will be proved in Section 7.

Remarks

The introduction of compound channels is due to BLACKWELL, BREIMAN, and THOMASIAN [2] and the author (WOLFOWITZ [4]), who independently proved Theorem 4.3.1. The proof given is that of the author, who also proved Theorem 4.4.1. (BLACKWELL, BREIMAN, and THOMASIAN proved the *weak* converse of Theorem 4.3.1.) The term "compound" is employed for the first time in the present monograph. It seems more appropriate than the terms used by all the above authors.

Section 4.5 is due to the author [4]. This latter paper also gives applications of these methods to other kinds of channels.

The introduction of the channels of Section 4.6 is due to SHANNON [3]. SHANNON proved the very interesting Theorem 4.6.4, and pointed out Theorem 4.6.3 and the results for channels I and II (with weak converses) as obvious. As in the preceding theorems, it is easy to indicate on what parameters the constants K and K' of Theorems 4.6.2 and 4.6.3 depend.

Theorem 4.8.1 is due to SHANNON [6].

Theorem 4.8.2 was proved independently by H. KESTEN and J. H. B. KEMPERMAN (oral communication).

The treatment of compound channels with feedback will be clear to the reader after a reading of Chapter 7.

5. The Discrete Finite-Memory Channel

5.1. The discrete channel. We begin this chapter with a description of the general discrete channel. The term "discrete" means that both input and output alphabets have a finite number of elements. For the sake of simplicity and without essential loss of generality we take both these alphabets to be $\{1, \ldots, a\}$. This will lighten somewhat the heavy burden of definitions and notation. The application of the theorems to cases where the alphabets have different numbers of elements will be obvious and will be made freely. (See also Chapter 3, Remark 10.) It is hoped that this manner of proceeding will make things less ponderous. Unless the contrary is explicitly stated the channels involved will have a single c.p.f., unlike, for example, the compound channels of Chapter 4.

A discrete channel is actually a system since it must be defined for each n, or at least for infinitely many n; for simplicity we will assume it defined for each n. Such a channel is simply the totality (for all n and u_0) of distributions of $v(u_0)$, i.e., the totality of values of

$$P_n\{v(u_0) = v_0\},$$

for every n and every pair u_0 and v_0 of n-sequences, where $v(u_0)$ is the chance received sequence when u_0 is sent over the channel. When it is not necessary to specify n explicitly to avoid possible ambiguity we shall write $P\{\ \}$ in place of $P_n\{\ \}$. Beyond the trivial requirement that

$$\sum_{v_0} P_n\{v(u_0) = v_0\} \equiv 1,$$

identically in n and u_0, where the summation is over all n-sequences v_0, no consistency conditions have to be met by the distributions. The definition of a code (N, λ) is obvious and we repeat it formally only for reference. Such a code is a system

$$\{(u_1, A_1), \ldots, (u_N, A_N)\} \tag{5.1.1}$$

where the u_i are n-sequences, the A_i are disjoint sets of n-sequences, and

$$P\{v(u_i) \varepsilon A_i\} \geqq 1 - \lambda, \quad i = 1, \ldots, N. \tag{5.1.2}$$

If we wish to indicate n explicitly we shall call it a code (n, N, λ).

We have already seen in Chapters 3 and 4 the (purely mathematical) utility of the introduction of the chance process (X, Y). We now proceed to extend and formalize this use of a chance process.

For any n consider the set D_n' of n-sequences of elements of the input alphabet and the set D_n'' of n-sequences of elements of the output alphabet as if they were formally different. (If one assumes, as we have, that the input and output alphabets are the same, then D_n' and D_n'' are identical. Otherwise one is a subset of the other.) Write D_n for the Cartesian product $(D_n' \times D_n'')$. Let D' and D'' be the (formally different) spaces of infinite sequences $(z_1, z_2, \ldots)$ in the elements of, respectively, the input and output alphabets, and let $D = (D' \times D'')$ be their Cartesian product. Any probability distribution Q_n' on D_n' will be called a "stochastic n-input" (sn-input). Thus a stochastic n-input Q_n' is any non-negative function defined on D_n' such that

$$\sum_{u_0} Q_n'(u_0) = 1,$$

where the summation is over all u_0 in D_n'. Any sequence Q_n', $n = 1, 2, \ldots$, of stochastic n-inputs will be called the "stochastic input sequence $\{Q_n'\}$". The Q_n' of a stochastic input sequence need satisfy no mutual consistency conditions whatever. Any probability distribution Q' on D'

will be called a "stochastic input" (s-input). For every n, Q' implies a marginal distribution Q'_n as follows: Let u_0 be any n-sequence in D'_n, and $S(u_0)$ be the cylinder set in D' with base u_0. Then $Q'_n(u_0) = Q'(S(u_0))$. A stochastic input sequence $\{Q'_n\}$, whose distributions are all marginal distributions of a stochastic input, of course fulfills obvious consistency conditions. In Chapters 3 and 4 we made use of a special sn-input, the distribution of X, implied by (2.1.6) and the fact that $X_1, \ldots, X_n$ were independent. When the latter is understood we can speak of the sn-input implied by π, and also of the s-input implied by π.

Let Q'_n be any sn-input. This, together with the totality of values of $P_n\{v(u_0) = v_0\}$, implies a distribution Q_n on D_n given by

$$Q_n(u_0, v_0) = Q'_n(u_0) \cdot P_n\{v(u_0) = v_0\},$$

called the joint sn-distribution. Q_n has a marginal distribution Q''_n on D''_n, called the stochastic n-output (sn-output) of the sn-input Q'_n (for a given channel), or, when Q'_n is understood, just the sn-output. Whenever an accented capital letter denotes an sn-input, the same letter accented twice denotes the sn-output of this sn-input, and the same letter without an accent denotes the corresponding distribution on D_n.

Let Q_n, Q'_n, Q''_n be as defined above. These are probability vectors and we may form their entropies as in Chapter 2, thus: $H(Q_n)$, $H(Q'_n)$, $H(Q''_n)$. $H(Q'_n)$ is called the entropy of the sn-input, $H(Q''_n)$ is called the entropy of the sn-output, and $H(Q_n)$ is called the joint entropy (of the sn-input and the (corresponding) sn-output). Of course, given the channel, all three are functions of Q'_n. The quantity

$$R_n(Q'_n) = H(Q'_n) + H(Q''_n) - H(Q_n)$$

is called the rate of transmission of the channel for the sn-input Q'_n, or briefly the rate for Q'_n. (Obviously R_n is a function of Q'_n.) For the d.m.c. and the sn-input implied by π, we have

$$R_n = nH(\pi) - n\sum_i \pi'_i H(w'(\cdot \mid i)) = nH(\pi') - n\sum_i \pi_i H(w(\cdot \mid i)).$$

Let Q_n be a probability distribution on D_n. Let (u, v) be a chance variable with values in D_n and distribution given by

$$P\{(u, v) = (u_0, v_0)\} = Q_n(u_0, v_0)$$

for any element (u_0, v_0) of D_n. Let I_n (the "mutual information function") be the function defined on D_n as follows: If $(u_0, v_0) \in D_n$ is such that $Q_n(u_0, v_0) = 0$ then $I_n(u_0, v_0) = 0$, and otherwise

$$I_n(u_0, v_0) = \log \frac{Q_n(u_0, v_0)}{Q'_n(u_0)\, Q''_n(v_0)}.$$

When we wish to put Q_n' in evidence we write $I_n(Q_n')$ and $I_n(u_0, v_0 \mid Q_n')$. Let J_n be the chance variable $I_n(u, v)$. If we wish to put Q_n' in evidence we write $J_n(Q_n')$. We have

$$E(J_n(Q_n')) = R_n(Q_n').$$

Finally, define formally for completeness

$$Q_n'(u_0 \mid v_0) = P\{u = u_0 \mid v = v_0\},$$

$$Q_n''(v_0 \mid u_0) = P\{v = v_0 \mid u = u_0\} = P\{v(u_0) = v_0\}.$$

Then the conditional entropy of the sn-output, given the sn-input, is defined to be

$$-E \log Q_n''(v \mid u) = \sum_{u_0 \in D_n'} Q_n'(u_0) H(Q_n''(\cdot \mid u_0)) = H(Q_n'' \mid Q_n') \quad \text{(say)},$$

and the conditional entropy of the sn-input, given the sn-output, is defined to be

$$-E \log Q_n'(u \mid v) = \sum_{v_0 \in D_n''} Q_n''(v_0) H(Q_n'(\cdot \mid v_0)) = H(Q_n' \mid Q_n'') \quad \text{(say)}.$$

From (2.2.2) we have

$$H(Q_n') + H(Q_n'' \mid Q_n') = H(Q_n) = H(Q_n'') + H(Q_n' \mid Q_n'').$$

Thus

$$H(Q_n') - H(Q_n' \mid Q_n'') = H(Q_n'') - H(Q_n'' \mid Q_n') = R_n(Q_n'). \quad (5.1.3)$$

For the d.m.c. and $Q_1' = \pi$, the capacity C is the maximum with respect to Q_1' of any member of (5.1.3) for $n = 1$.

5.2. The discrete finite-memory channel. It would now seem to be a good idea to impose on the general channel of the preceding section restrictions which would be realistic and (we hope) lead to interesting results. Such conditions will obviously involve, inter alia, consistency requirements among the elements of the totality $\{P_n\{v(u_0) = v_0\}\}$ which is the channel. There is obviously no limit on the variety of restrictions which could be imposed. This section is devoted to the study of what has been called "the discrete finite-memory channel" (d.f.m.c.). The reasons for the importance of this channel will be discussed later. Our treatment will differ from that usually given in the literature, and the reasons for the difference will be explained later.

We need some new notation. If $s = (s_1, \ldots, s_n)$ is an n-sequence in D_n' or D_n'', let $s(c, d) = (s_c, \ldots, s_d)$. Thus $s(c, d)$ is a sequence whose elements are coordinates of index $c, c+1, \ldots, d$, respectively. When this sequence is moved so that its elements are coordinates of index $g, \ldots, (d-c+g)$, respectively, we write $s^{(g)}(c, d)$. Let $s(c, d, D_n')$ ($s(c, d, D_n'')$) denote the cylinder set in D_n' (in D_n'') with base $s(c, d)$. Let $s^{(g)}(c, d, D_n')$ and $s^{(g)}(c, d, D_n'')$ be the corresponding cylinder sets over $s^{(g)}(c, d)$.

A discrete channel is said to be a d.f.m.c. if there exists a non-negative integer m, called "the duration of memory", such that the following two conditions are satisfied for every positive integral n, and every $u_0 \varepsilon D'_n$ and $v_0 \varepsilon D''_n$:

If
$$m+1 \leqq c \leqq d \leqq n,$$
then
$$P_n\{v(u_0) \varepsilon v_0(c, d, D''_n)\}$$
$$= P_{d-c+m+1}\{v(u_0^{(1)}(c-m, d)) \varepsilon v_0^{(m+1)}(c, d, D''_{d-c+m+1})\}. \tag{5.2.1}$$
If
$$1 \leqq c \leqq d, d+m+1 \leqq g \leqq h \leqq n,$$
then
$$P_n\{v(u_0) \varepsilon [v_0(c, d, D''_n) \cap v_0(g, h, D''_n)]\} \tag{5.2.2}$$
$$= P_n\{v(u_0) \varepsilon v_0(c, d, D''_n)\} \cdot P_n\{v(u_0) \varepsilon v_0(g, h, D''_n)\}.$$

The first of these conditions says essentially that the probability of a finite received sequence depends only on the corresponding symbols transmitted and on the m preceding transmitted symbols. The second condition says essentially that two received sequences separated by at least m symbols are independently distributed. If a d.f.m.c. has duration of memory m, it also has duration of memory $m' > m$. Usually it is the smallest integer with this property which is called the duration of memory.

Let l be any positive integer. By the d.m.c. $(D'_{l+m}, D''_l, P_{l+m})$ we shall denote the d.m.c. whose input *alphabet* consists of the sequences of D'_{l+m}, whose output *alphabet* consists of the sequences of D''_l, and whose c.p.f. $w(\cdot \mid \cdot)$ is defined as follows: Let $u_0 \varepsilon D'_{l+m}$, $v_0 \varepsilon D''_l$. Let $v_0^* \varepsilon D''_{l+m}$ be any sequence such that $v_0^{*(1)}(m+1, m+l) = v_0$. Then
$$w(v_0 \mid u_0) = P_{l+m}\{v(u_0) \varepsilon v_0^*(m+1, m+l, D''_{l+m})\}. \tag{5.2.3}$$
Let $C^*(l+m)$ be the capacity of this channel. We now define
$$C = \sup_l \frac{C^*(l+m)}{(l+m)}. \tag{5.2.4}$$
Since $C^*(l+m) \leqq (l+m)\log a$, $C \leqq \log a$. It will soon appear that C is the capacity of the d.f.m.c., and it is a consequence of Theorem 5.5.1 below that
$$C = \lim_{l\to\infty} \frac{C^*(l+m)}{(l+m)}. \tag{5.2.5}$$

5.3. The coding theorem for the d.f.m.c. The following coding theorem is very easy to prove by reduction to the discrete memoryless case.

Theorem 5.3.1. Let $\varepsilon > 0$ and λ, $0 < \lambda \leqq 1$, be arbitrary. For any d.f.m.c. and all n sufficiently large there exists a code $(n, 2^{n(C-\varepsilon)}, \lambda)$.

Proof: Let l be a positive integer such that

$$\frac{C^*(l+m)}{(l+m)} > C - \frac{\varepsilon}{2}. \tag{5.3.1}$$

Let $n = k(l+m)$, with k an integer. If v_0 is any n-sequence let $\bar{v}_0$ denote the subsequence of v_0 whose elements are those coordinates of v_0 whose indices are of the form

$$k'(l+m)+j, \quad k' = 0, \ldots, k-1; \quad j = (m+1), \ldots, (m+l)$$

(i.e., congruent to $m+1$, $m+2, \ldots, m+l$, modulo $(l+m)$). (We shall have frequent occasion in this chapter to use this operation on a sequence, e.g., $v(u_0)$ and $\bar{v}(u_0)$.) Let $B(\bar{v}_0)$ be the cylinder set in D''_n with base $\bar{v}_0$. Each such cylinder set obviously contains a^{km} n-sequences.

Let u_0 be any n-sequence $(n = k(l+m))$ in D'_n. Then u_0 is a k-sequence in the input alphabet of the d.m.c. $(D'_{(l+m)}, D''_l, P_{(l+m)})$. It follows from (5.2.1) and (5.2.2) that, for any u_0 and v_0,

$$P_n\{v(u_0) \varepsilon B(\bar{v}_0)\} = P_n\{\bar{v}(u_0) = \bar{v}_0\}$$

is the probability that, when u_0 is sent over the d.m.c. $(D'_{(l+m)}, D''_l, P_{(l+m)})$, the sequence v_0 will be received. Consequently a code (k, N, λ) for the d.m.c. $(D'_{(l+m)}, D''_l, P_{(l+m)})$ is a code $(k(l+m), N, \lambda)$ for the original d.f.m.c. (but not vice versa). Now we have, by Theorem 3.2.1, that there exists, for all k sufficiently large, a code (k, N, λ) for the above d.m.c., with

$$N > 2^{k\left(C^*(l+m)-\frac{\varepsilon}{4}\right)} = 2^{n\left(\frac{C^*(l+m)}{(l+m)}-\frac{\varepsilon}{4(l+m)}\right)} > 2^{n\left(C-\frac{3\varepsilon}{4}\right)}. \tag{5.3.2}$$

This proves the theorem for n of the form $k(l+m)$.

Suppose now $n = k(l+m)+t$, $1 \leqq t < (l+m)$. Proceed exactly as before. (5.3.2) now can be written as

$$N > 2^{(n-t)\left(C-\frac{3\varepsilon}{4}\right)} = 2^{n\left(\frac{n-t}{n}\right)\left(C-\frac{3\varepsilon}{4}\right)} > 2^{n(C-\varepsilon)} \tag{5.3.3}$$

for k (or n) sufficiently large. This completes the proof of the theorem.

It follows from (3.2.10) or Section 7.5 below that λ in the statement of Theorem 5.3.1 could be replaced by an expression of the form $c_1 e^{-nc_2}$, where c_1 and c_2 are properly chosen positive constants.

5.4. Strong converse of the coding theorem for the d.f.m.c.

Theorem 5.4.1. Let $\varepsilon > 0$ and λ, $0 \leq \lambda < 1$, be arbitrary. For a d.f.m.c. and all n sufficiently large, any code (n, N, λ) must satisfy

$$N < 2^{n(C+\varepsilon)}.$$

Proof: We shall prove the theorem for $\lambda > 0$; the theorem is then a fortiori true when $\lambda = 0$. Let l be a positive integer to be chosen later. We shall first prove the theorem for n of the form $k(l+m)$, with k a sufficiently large integer, and then for all n sufficiently large.

Suppose

$$\{(u_1, A_1), \ldots, (u_N, A_N)\}$$

is a code (n, N, λ) for a d.f.m.c. Let v_0 be any n-sequence in A_i $(i=1,\ldots,N)$, and let $\bar{v}_0$ and $B(\bar{v}_0)$ be as in Section 5.3. Let $\bar{A}_i$ be the set of $\bar{v}_0$ for all v_0 in A_i. Define

$$A_i^* = \cup\, B(\bar{v}_0) \tag{5.4.1}$$

where the union is over all v_0 in A_i. The A_i^* are not necessarily disjoint. However, a fortiori,

$$P_n\{v(u_i)\,\varepsilon A_i^*\} = P_n\{\bar{v}(u_i)\,\varepsilon \bar{A}_i\} \geqq 1-\lambda,\ i = 1, \ldots, N. \tag{5.4.2}$$

It follows from (5.2.1) and (5.2.2) that the middle member of (5.4.2) is also the probability that, when u_i is sent over the d.m.c. $(D'_{(l+m)}, D''_l, P_{(l+m)})$, the sequence received lies in $\bar{A}_i$.

Let λ' be such that $\lambda < \lambda' < 1$. By choosing the δ of (2.1.3) sufficiently large (for the d.m.c. $(D'_{l+m}, D''_l, P_{l+m})$) we can achieve the following: Delete from $\bar{A}_i$ all kl-sequences not generated by u_i with respect to the c.p.f. of the d.m.c. $(D'_{l+m}, D''_l, P_{l+m})$; call the set after deletion $\bar{A}_i^*$. Then

$$P_n\{\bar{v}(u_i)\,\varepsilon \bar{A}_i^*\} \geqq 1-\lambda',\ i = 1, \ldots, N. \tag{5.4.3}$$

The set

$$\{(u_1, \bar{A}_1^*), \ldots, (u_N, \bar{A}_N^*)\} \tag{5.4.4}$$

would be a code (k, N, λ') for the d.m.c. $(D'_{l+m}, D''_l, P_{l+m})$ were it not for the fact that the $\bar{A}_i^*$ are not necessarily disjoint. However, any $\bar{v}_0$ can belong to at most a^{km} sets $\bar{A}_i^*$. Consequently an obvious modification of the proof of Theorem 3.3.1 enables us to conclude that, for k sufficiently large,

$$\begin{aligned} N &< a^{km} \cdot 2^{k\left(C^*(l+m)+\frac{\varepsilon}{4}\right)} < a^{n\cdot\frac{m}{(m+l)}} \cdot 2^{n\left(\frac{C^*(l+m)}{l+m}+\frac{\varepsilon}{4}\right)} \\ &< a^{n\cdot\frac{m}{(m+l)}} \cdot 2^{n\left(C+\frac{\varepsilon}{4}\right)}. \end{aligned} \tag{5.4.5}$$

Now choose l so large that $\frac{m \log a}{m+l} < \frac{\varepsilon}{4}$, and the theorem is proved for all sufficiently large n of the form $k(l+m)$.

Finally, suppose $n = k(l+m)+t$, with k an integer and $1 \leqq t < l+m$. Then, writing $n' = (k+1)(l+m)$, we have from (5.4.5) that

$$N < 2^{n'\left(C+\frac{\varepsilon}{2}\right)} < 2^{n\left(1+\frac{l+m}{n}\right)\left(C+\frac{\varepsilon}{2}\right)} < 2^{n(C+\varepsilon)} \tag{5.4.6}$$

for n sufficiently large. This proves the theorem.

5.5. Rapidity of approach to C in the d.f.m.c. Let l be any positive integer and $\varepsilon > 0$ be arbitrary. Let $n = k(l+m)$. From Theorem 5.3.1 and inequality (5.4.5) we obtain that, for k sufficiently large

$$2^{n(C-\varepsilon)} < 2^{n\left(\frac{m \log a}{m+l} + \frac{C^*(l+m)}{l+m} + \frac{\varepsilon}{4}\right)}$$

so that

$$C - \varepsilon < \frac{m \log a}{m+l} + \frac{C^*(l+m)}{l+m} + \frac{\varepsilon}{4}. \tag{5.5.1}$$

Since ε was arbitrary (5.5.1) could not hold unless

$$C \leqq \frac{m \log a}{m+l} + \frac{C^*(l+m)}{l+m}.$$

From the definition of C we have

$$C \geqq \frac{C^*(l+m)}{l+m}.$$

Thus we have

Theorem 5.5.1. For any d.f.m.c., and any l, we have

$$\frac{C^*(m+l)}{(m+l)} \leqq C \leqq \frac{C^*(m+l)}{(m+l)} + \frac{m \log a}{(m+l)} \tag{5.5.2}$$

and (5.2.5) holds. C is zero if and only if $C^*(m+l) = 0$ for every positive integer l.

This result enables us, at least in principle, to compute the capacity C of a d.f.m.c. to any desired degree of approximation, since $C^*(l+m)$ is in principle computable to any desired degree of approximation.

5.6. Discussion of the d.f.m.c. The d.f.m.c. has an obvious and natural importance. Channels with memory, i.e., where the distribution of a letter in the received message depends upon previous letters sent (and/or received) obviously offer practical and mathematical interest. The d.f.m.c. is complex enough to fit many practical situations and yet simple enough so one can come to mathematical grips with it. It is free of a certain difficulty which arises in the continuous operation of a channel and which will be described in the next chapter. On the other hand, there are many channels of at least mathematical, and perhaps even of practical interest, in which the received message depends upon previous *received* messages in a manner which does not satisfy condition (5.2.2). Moreover, it would be reasonable for the "memory" of preceding transmitted letters to decline with the elapsed "time" but not to cease abruptly after m letters. These remarks are not intended to be precise, and cannot be made precise in the required generality except after the introduction of the proper mathematical model. After the latter is done it becomes appropriate to add additional restrictions to the definition of a channel beyond those given in Section 5.1. This will be done in the next chapter, after some preliminary discussion to motivate the restrictions.

Suppose the n-sequence u_0 is sent over a d.f.m.c., and let

$$(Y_1(u_0), \ldots, Y_n(u_0))$$

be the chance received sequence. If m is the duration of memory, then obviously $Y_{m+1}(u_0)$ and $Y_{2m+2}(u_0)$, say, are independently distributed. However, it would be an error to conclude that, when $Y_{m+1}(u_0), \ldots,$ $Y_{2m+1}(u_0)$ are given, the distribution of $Y_{2m+2}(u_0)$ does not depend on the value of $Y_{m+1}(u_0)$. Thus, the chance variables $Y_i(u_0)$ do not, in general, constitute a Markov chain, even one of order m.

A special case of the d.f.m.c. occurs when, for every u_0, $Y_1(u_0), \ldots,$ $Y_n(u_0)$ are independently distributed. For lack of a better name we will call such a channel a discrete independent finite memory channel (d.i.f.m.c.). Suppose given such a channel. Let u be a *chance* sequence, with values in D'_n, whose elements are independent, identically distributed chance variables. It would be an error to conclude that $Y_1(u), \ldots,$ $Y_n(u)$ are independently distributed. Suppose u is as before but with values in D' (definition in Section 5.1). It is not necessarily true that $Y_1(u), Y_2(u), \ldots$ constitute a Markov chain of any finite order.

Suppose given a compound d.f.m.c., all of whose c.p.f.'s have the same duration of memory m. (The latter statement means only that there is a number m which can serve as the duration of memory for all the c.p.f.'s. It is not precluded that the smallest number which can serve as the duration of memory of any c.p.f. should be a function of the c.p.f.) Now the d.f.m.c. was treated by a reduction to the d.m.c., and this reduction can be effected uniformly for all the c.p.f.'s. The resulting memoryless channels can be treated by the methods of Chapter 4. The details are routine and left to the reader.

The treatment of the d.f.m.c. in the literature differs considerably from that of the present chapter. (For an excellent description of the customary treatment see FEINSTEIN [2], Ch. 6.) The now conventional treatment takes as its point of departure the space D of Section 5.1 (more properly, a space analogous to D but consisting of doubly-infinite sequences. See Section 6.2 below.) A stationary stochastic input μ' on D' then implies a stationary distribution μ on D. A major concern is then to prove the ergodicity of μ, given the ergodicity of μ'. (See Section 10.3 below.) The ergodicity of μ is then used to prove the "asymptotic equipartition property", i.e., that $\frac{1}{n} J_n(\mu')$ converges, with probability one or in L^1, to

$$\lim_{n\to\infty} \frac{1}{n} R_n(\mu'_n), \tag{5.6.1}$$

which is shown to exist. (See Section 10.2 below.) The asymptotic equipartition property then gives a coding theorem, by means of Theorem 7.2.1 or other argument. (Convergence in probability suffices for Theorem 7.2.1 to give a coding theorem.)

It is often the practice in the literature to call the supremum C_e (C_s) of (5.6.1) with respect to all stationary ergodic μ' (with respect to all stationary μ') the "ergodic capacity" (the "stationary capacity") of the channel. The capacity of a channel (when the channel has a "capacity" which is the same for all λ between 0 and 1) is an intrinsic characteristic of the channel. Its existence and actual value have to be demonstrated by proving a coding theorem and a strong converse, and are not matters merely of definition.

The problem has been posed, as one of great importance to the study of the d.f.m.c., whether $C_e = C_s$. This question has been resolved in the affirmative in several ingenious proofs by different writers. In two of these recently published proofs it is proved that $C_e = C_s$ by proving that $C_e = C$ and $C_s = C$, where C is the quantity defined in (5.2.4).

The view to which we adhere throughout this monograph is that in coding theory stochastic inputs serve only as a tool in proving coding theorems and are not of interest per se. (Whether the study of stochastic inputs is of mathematical interest outside of coding theory is not relevant here and need not be decided by us.) In the present chapter we have proved a coding theorem and a strong converse for the d.f.m.c., using proofs of great simplicity. We showed that the capacity of a d.f.m.c. could be computed to any desired degree of accuracy by computing the rate of transmission for a suitable stochastic input of independent, identically distributed sequences (of length $(l + m)$). Whether the then resulting μ is ergodic and whether $C_e = C_s$ are questions which did not arise.

Remarks

What has been called above the "duration of memory" is called in the literature (including papers of the author) the "memory". The reason for the present change of name is a desire to avoid the grammatical barbarisms and the confusion which would result from using the word "memory", both in its original meaning and in the sense of duration of memory, in the same or in adjacent sentences.

The strong converse (Theorem 5.4.1) for the d.i.f.m.c. was proved independently by FEINSTEIN [*3*] and the author (WOLFOWITZ [*2*]). Theorems 5.4.1 and 5.5.1 are due to the author (WOLFOWITZ [*5*]).

6. Channels with Arbitrarily Varying Channel Probability Functions

6.1. Introduction. Let S be a finite collection of c.p.f.'s as described in Section 4.1. Suppose that s varies arbitrarily *from letter to letter*, instead of remaining constant during the transmission of a word of length n. Let $S^t = S$, $t = 1, \ldots, n$. For every n-sequence $s_n = (s^1, \ldots, s^n) \varepsilon \prod_{t=1}^{n} S^t = S_n$ we define

$$P_{s_n}\{v(u_0) = v_0\} = P\{v(u_0) = v_0 | s_n\}$$
$$= \prod_{t=1}^{n} w(y_t | x_t | s^t) . \qquad (6.1.1)$$

Such a channel is called "arbitrarily varying". Results on such channels have not yet reached the level of results on compound channels because the problems are much more difficult. This chapter is devoted to some of the results already obtained.

6.2. Necessary and sufficient conditions for a positive rate of transmission. Let a and b, respectively, be the sizes of the input and output alphabets. We assume that S contains c c.p.f.'s $w(\cdot|\cdot|i)$, $i = 1, \ldots, c$. For any j, $j = 1, \ldots, a$, let $T(j)$ be the convex hull of the c points of b-space,

$$(w(1|j|i), w(2|j|i), \ldots, w(b|j|i)), \; i = 1, \ldots, c .$$

We shall now prove

Theorem 6.2.1. Necessary and sufficient for the possibility of a positive rate of transmission is that at least two of the convex bodies $T(1), \ldots, T(a)$ be disjoint. (A positive rate of transmission is said to be possible if, for any λ $(0 < \lambda < 1)$ and n sufficiently large, there exists a code of length $\exp_2\{n m\}$, $m > 0$.)

Proof of necessity: Suppose that no two of the convex bodies are disjoint. Fix n and $\lambda < 1/2$. We shall show that, for any code (n, N, λ), $N = 1$. Suppose the code is $(u_1, A_1), \ldots, (u_N, A_N)$.

We can think of s_n as being chosen by an individual called "the jammer". The jammer can choose s_n at random, according to a distribution δ, say, on s_n. We define

$$P\{v(u_0) \varepsilon A | \delta\} = \sum_{S_n} P\{v(u_0) \varepsilon A | s_n\} \delta(s_n) . \qquad (6.2.1)$$

We shall construct a δ such that

$$P\{v(u_1) \varepsilon A_2 | \delta\} \geqq 1 - \lambda . \qquad (6.2.2)$$

Obviously, from the definition of a code,

$$P\{v(u_1) \varepsilon A_1 | \delta\} \geqq 1 - \lambda . \qquad (6.2.3)$$

Since $\lambda < 1/2$ and A_1 and A_2 are disjoint, there is a contradiction between (6.2.2) and (6.2.3).

Suppose $u_1 = (x_1, \ldots, x_n)$ and $u_2 = (y_1, \ldots, y_n)$. Let z_1, $i = 1, \ldots, n$, be a point common to $T(x_i)$ and $T(y_i)$. Let $g_i = (g_1^{(i)}, \ldots, g_c^{(i)})$ and $h_i = (h_1^{(i)}, \ldots, h_c^{(i)})$ be the barycentric coordinates of z_i in the sets $T(x_i)$ and $T(y_i)$, respectively. Actually, each set T can have at most min (b, c) extreme points, and therefore each point in T need have at most that many non-zero barycentric coordinates. The barycentric coordinates corresponding to interior points should be set equal to zero.

Now let δ (resp. δ^*) be a distribution on S_n implied by the following: The elements of s_n are independent chance variables, the distribution of the ith chance variable being g_i (resp, h_i). It follows that

$$P\{v(u_1) \varepsilon A_2 | \delta\} = P\{v(u_2) \varepsilon A_2 | \delta^*\}. \tag{6.2.4}$$

The right member of the above is $\geqq 1 - \lambda$ by the definition of a code. This implies (6.2.2) and proves the necessity of the condition.

Proof of sufficiency: Without loss of generality we may assume that $T(1)$ and $T(2)$ are disjoint. Then there is a plane in b-space which separates $T(1)$ and $T(2)$ and is disjoint from both; let $(l_1, \ldots, l_b, m)$ be its coordinates. Suppose a q-sequence z_1 (resp. z_2) consisting entirely of ones (resp., of twos) is sent over the channel. Choose arbitrarily η, $0 < \eta < 1/2$. Let $N_i(z_j)$, $i = 1, \ldots, b$; $j = 1, 2$, be the number of elements i in the chance sequence $v(z_j)$. Choosing q sufficiently large and reversing, if necessary, the indices 1 and 2, we conclude from the law of large numbers that always

$$P\left\{\sum_{i=1}^{b} l_i N_i(z_1) < qm\right\} > 1 - \eta \tag{6.2.5}$$

and

$$P\left\{\sum_{i=1}^{b} l_i N_i(z_2) \geqq qm\right\} > 1 - \eta, \tag{6.2.6}$$

no matter what s_n is chosen by the jammer. We now show that it is possible to achieve a positive rate of transmission by using as letters only q-sequences of ones and twos.

Let $C(\eta)$ be the capacity of a binary symmetric channel with c.p.f. w as follows:

$$w(1/1) = w(2|2) = 1 - \eta$$
$$w(2/1) = w(1|2) = \eta$$

$C(\eta)$ is positive. By (6.2.5) and (6.2.6) the receiver can correctly decode transmitted q-sequences of ones and twos with probability at least $1 - \eta$, no matter what s_n may be. Hence, by Theorem 3.2.1, for any λ, $0 < \lambda < 1$, there exists a code of length at least

$$\exp_2\left\{\frac{n}{q} C(\eta) - K\sqrt{\frac{n}{q}}\right\}.$$

This proves the sufficiency part of the theorem.

We now prove

Theorem 6.2.2. Suppose the receiver but not the sender knows the *actual* channel sequence s_n used in the transmission of a word. Then, necessary and sufficient for the possibility of a positive rate of transmission is that, for some two letters of the input alphabet, say d_1 and d_2,

$$\sum_{i=1}^{b} |w(i|d_1|j) - w(i|d_2|j)| > 0, j = 1, \ldots, c. \tag{6.2.7}$$

Proof of necessity: Suppose (6.2.7) does not hold. Then, for any two letters of the input alphabet, there exists a value j such that, under the jth c.p.f., the distribution fo the letters received is the same for these two letters of the input alphabet. Consequently, for u_1 and u_2, for example, there exists a sequence s_n such that the distribution of $v(u_1)$ and $v(u_2)$ is the same, when s_n governs the transmission of either sequence. If $\lambda < 1/2$ this leads to essentially the same contradiction as in the proof of necessity of Theorem 6.2.1.

Proof of sufficiency: This is only slightly different from the proof of sufficiency of Theorem 6.2.1. Without loss of generality we may take $d_1 = 1$, $d_2 = 2$. For q sufficiently large and any fixed j, $j = 1, \ldots, c$, the receiver can decide whether a q-sequence of ones or of twos was sent by the sender, with probability of error at most η, if he knows the j (i.e., the c.p.f.) being employed. This follows from (6.2.7) and the law of large numbers exactly as in the proof of Theorem 6.2.1. We now proceed as in the proof of sufficiency of this latter theorem, except that we send cq-sequences of ones or twos instead of q-sequences. In each cq-sequence transmitted, there must be at least q letters transmitted according to the same $w(\cdot|\cdot|\cdot)$. The receiver knows each w and uses the first q letters transmitted according to the same w to decode the letter (cq-sequence). The conclusion is then the same as in Theorem 6.2.1, except that q is replaced by cq. Actually it would be enough to use sequences of length $c(q-1)+1$.

For $k = 1, \ldots, c$, let $D(k)$ be the set of a points in b-space

$$\{D(i|k) = (w(1|i|k), \ldots, w(b|i|k), i = 1, \ldots, a\}.$$

Let $B'(1), \ldots, B'(a^c)$ be sets, each of which contains c points, one from each of $D(1), \ldots, D(c)$. Let $B(i)$ be the convex hull of $B'(i)$.

The previous theorem dealt with the case where the receiver but not the sender always knows the *actual* s_n which governed the transmission of a word. We want now to deal with the case where the sender knows s_n and the receiver does not. The results will be divided into two theorems, of which the second, which deals with the necessity condition, is weaker than the necessity part of Theorem 6.2.2.

Theorem 6.2.3. Sufficient for the possibility of a positive rate of transmission, when the sender but not the receiver knows the sequence s_n used in the transmission of a word, is that at least two of the $B(i)$ be disjoint.

Suppose $B(1)$ and $B(2)$ are disjoint. The sender's words in Theorem 6.2.1 are q-sequences of ones and twos. Now the idea is the same, except that, when the sender wishes to send a one (resp., a two) and knows that $w(\cdot|\cdot|k)$ will govern its transmission, he sends instead the letter $a_1(k)$ such that $D(a_1(k)|k)$ (resp., the letter $a_2(k)$ such that $D(a_2(k)|k)$ is contained in $B(1)$ (resp., $B(2)$). The rest of the proof essentially follows the idea of the proof of sufficiency of Theorem 6.2.1.

Theorem 6.2.4. Suppose that the jammer is allowed to choose the sequence s_n which governs the transmission of a word according to a probability distribution δ, say. The sender knows δ but not s_n, the actual outcome of the chance experiment. The receiver does not know s_n. If no two of the $B(i)$ are disjoint then it is impossible to transmit over the channel at a positive rate.

The proof is essentially the same as the proof of necessity of Theorem 6.2.1, bearing in mind the changes in proof made in Theorem 6.2.3.

Let $C(i)$ be the capacity of the d.m.c. with c.p.f. $w(\cdot|\cdot|i)$. Let $C^* = \min C(i)$. Suppose both sender and receiver know the s_n which governs the transmission of a word. It does not matter, for our result below, whether or not the sender knows the entire sequence s_n in advance, or only element by element as each letter comes up for transmission. As for the receiver, he decodes only after the entire sequence has been received, so that this distinction does not arise. It is obvious that the capacity of this channel cannot be greater than C^*. We shall prove

Theorem 6.2.5. The capacity of the channel, where both sender and receiver know the s_n which governs the transmission of a word, is C^*.

Proof: Clearly it is sufficient to prove that C^* is a possible rate of transmission. For typographical simplicity we shall always write $\sqrt{n}$ as if it were an integer and leave to the reader the obvious and trivial changes to be made if it is not.

We recall to the reader the result (3.2.10). Suppose s_n has the following structure: The first $\sqrt{n}$ elements are all d_1, the next $\sqrt{n}$ elements are all d_2, etc., so that s_n is determined by $(d_1, \ldots, d_{\sqrt{n}})$. Both sender and receiver know this last sequence. For each element d_i of this sequence we can construct a code

$$(\sqrt{n}, \exp_2\{\sqrt{n}(C^* - \varepsilon)\}, c_1 \exp_e\{-\sqrt{n}c_2\}) .$$

Combining these codes in the obvious way we have a code, for the original channel with known s_n,

$$(n, \exp_2\{n(C^* - \varepsilon)\}, c_1 \sqrt{n} \exp_e\{-\sqrt{n}\, c_2\}) .$$

When n is sufficiently large the probability of error can be made as small as desired.

It remains only to discuss the case when s_n does not have the above structure. Recall that both sender and receiver know s_n. Sender and receiver agree in advance upon a procedure Π, say, which will rearrange *any* sequence s_n into a unique one which has the structure of the last paragraph, except possibly at its end; we will discuss this in a moment. It is easy to envision such a procedure. Let $\Pi^{-1}(s_n)$ be its inverse, so that $\Pi^{-1}(s_n) \circ \Pi \circ s_n \equiv s_n$. The sender applies $\Pi^{-1}(s_n)$ to the code word he would send if $\Pi \circ s_n$ were the sequence of c.p.f.'s which governs the transmission of the word, and sends the resulting word. The receiver applies Π to the sequence he receives, and then decodes the resulting sequence as if it had been transmitted over the channel with $\Pi \circ s_n$ as the sequence of c.p.f.'s to govern the transmission of the word. This establishes the theorem except for the difficulty mentioned earlier, which might occur at the end. We will now discuss this difficulty.

When any sequence s_n is rearranged into sequences of $\sqrt{n}$ identical elements, there can be at most $c(\sqrt{n} - 1)$ elements which cannot be arranged into $\sqrt{n}$-sequences of identical elements. Consequently there will be at least $n - c(\sqrt{n} - 1)$ elements which can be arranged into $\sqrt{n}$-sequences of identical elements. These form at least $(\sqrt{n} - c)$ sequences of length $\sqrt{n}$. We therefore have Π rearrange s_n in a unique way into $(\sqrt{n} - c)$ consecutive sequences of length $\sqrt{n}$, and "ignore" the remaining elements of s_n, which may conveniently be put at the end of the rearranged sequence. The number of elements so ignored is $n - \sqrt{n}(\sqrt{n} - c) = c\sqrt{n}$. Similarly we ignore the last $c\sqrt{n}$ elements of the properly rearranged transmitted word and received word. Clearly this does not affect the capacity, and the theorem is proved.

6.3. Remarks on the capacity of an arbitrarily varying channel. It is tempting to conjecture, and it has actually been asserted in print, that the capacity C of an arbitrarily varying channel is C', the minimum of the capacities of all matrices (c.p.f.'s) whose j^{th} row is an element of $T(j)$ (defined in the preceding section) for every $j = 1, \ldots, a$. Call this set of matrices $\bar{S}$. Implicit in the idea of this conjecture is that any letter sent can be chosen according to a probability distribution on the input letters $1, \ldots, a$, independently of all other letters sent. We make this assumption explicitly. It is also made in the same place where the assertion is stated as a theorem. We will now show, by an example, that this conjecture is not true.

Suppose $a = b = 3$, $c = 2$. Let w and w' be the two matrices. Denote the rows of w (resp., w') by 1, 2, 3 (resp., 1', 2', 3'). Each row can be denoted by a point in the plane of its first two elements, since the three elements add to one. Suppose the resulting diagram looks as follows:

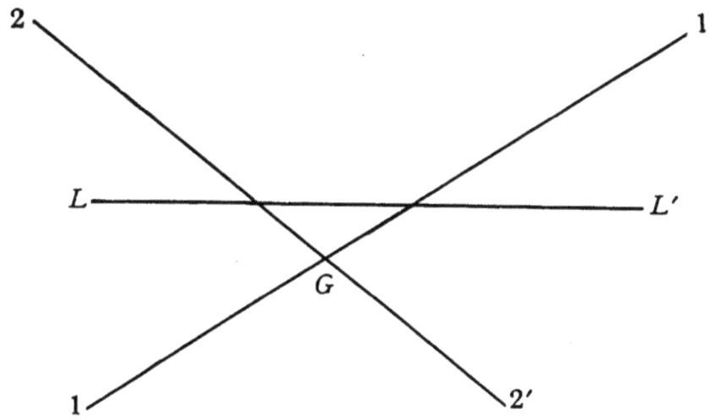

Here $3 = 3' = G$, the point of intersection of the two lines. L (resp., L') is the midpoint of the line 1—2 (resp., $1' - 2'$). First we show that $C' = 0$. Since G is the intersection of the two lines drawn, G is in $T(1)$ and in $T(2)$. It is also the sole member of $T(3)$. Consequently, the matrix, all of whose elements are G, is in $\bar{S}$, and its capacity is zero. Thus the conjecture requires that $C = 0$.

We will now show the possibility of a positive rate of transmission, which contradicts the conjecture that $C = 0$. The sender now uses only two letters. One is the old letter 3. The other letter, to be called 4, say, consists in sending 1 with probability 1/2, and 2 with probability 1/2. Thus $T(3) = G$, as before, and $T(4)$ is the line LL'. Since these are disjoint, it is possible to transmit at a positive rate, by Theorem 6.2.1.

In the next section we will prove that the conjecture described in this section is valid when $b = 2$, even when S contains infinitely many elements.

6.4. The capacity C of an arbitrarily varying channel when b = 2.

In this section we prove

Theorem 6.4.1. When $b = 2$, $C = C'$. S may have infinitely many elements.

The theorem also holds when the sender may randomize each letter or, indeed, the entire word to be sent. We leave the simple proof of this to the reader, and henceforth proceed in this section without permitting randomization to the sender.

Let $X = \{1,\ldots, a\}$, $X^t \equiv X$, $Y = \{1, 2\}$, $Y^t \equiv Y$, $X_n = \prod_{t=1}^{n} X^t$, $Y_n = \prod_{t=1}^{n} Y^t$. Write $y_n \varepsilon Y_n$ as $(y^1,\ldots, y^n)$. This represents a change of notation from previous chapters, which is regretable but essential. Let $\overline{w}$ be the matrix whose capacity is C'. By choosing each element of s_n independently according to a suitable probability distribution, the jammer can achieve that transmission of each word is governed by the matrix $\overline{w}$. Hence the capacity, if it exists, cannot exceed C'. We now show that $C = C'$.

In Lemma 6.4.1 we shall assume that there are only two matrices, say w and w'. (This is the only lemma where we make this restriction.) Let the rows of w (resp., w') be denoted temporarily by 1 and 2 (resp., 1' and 2'). The rows can represented as points on the line $x + y = 1$ in the (x, y)-plane. Let the abscissa of the point be the first element of the row. Suppose first, until further notice, that the points of w and w' fall in the following configuration:

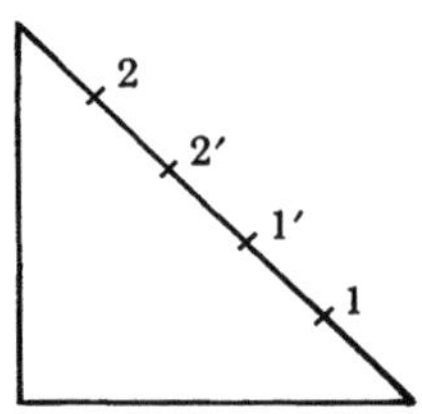

We shall say that the code $(u_1, A_1),\ldots, (u_N, A_N)$ is a strict maximum likelihood code (s.m.l.c.) with respect to w' if, for $i = 1,\ldots, N$,

$$A_i = \{y_n | y_n \varepsilon Y_n \text{ and } P\{v(u_i) = y_n | w'\} > P\{v(u_j) = y_n | w'\} \text{ for } j \neq i\}. \tag{6.4.1}$$

Write $u_i = (u_i^1,\ldots, u_i^n)$. Define

$${}_1A_i^t = \{y_n | y_n \varepsilon A_i \quad \text{and} \quad y^t = u_i^t\}$$

$${}_2A_i^t = \{y_n | y_n \varepsilon A_i \quad \text{and} \quad y^t \neq u_i^t\}$$

$${}_1A_i^{*t} = \{(y^1,\ldots, y^{t-1}, y^{t+1},\ldots, y^n) | \text{ there exists } y^t \text{ such that } (y^1,\ldots, y^n) \varepsilon {}_1A_i^t\}$$

$${}_2A_i^{*t} = \{(y^1,\ldots, y^{t-1}, y^{t+1},\ldots, y^n) | \text{ there exists } y^t \text{ such that } (y^1,\ldots, y^n) \varepsilon {}_2A_i^t\}.$$

Recall that we are still in the case of the configuration of the last diagram. We now prove

Lemma 6.4.1. With w and w' as in the figure, if $\{(u_1, A_1), \ldots, (u_N, A_N)\}$ is an s.m.l.c. for w', then

$$_1A_i^{*t} \supseteq {}_2A_i^{*t},\ i = 1, \ldots, N;\ t = 1, \ldots, n \tag{6.4.2}$$

and, if $s_n^* = (w', w', \ldots, w')$, then

$$P\{v(u_i)\,\varepsilon A_i \mid s_n\} \geqq P\{v(u_i)\,\varepsilon A_i \mid s_n^*\} \tag{6.4.3}$$

for $i = 1, \ldots, N$ and all $s_n \,\varepsilon\, S_n$.

Proof: Suppose first that $u_i^t = 1$. Let

$$(y^1, \ldots, y^{t-1}, y^{t+1}, \ldots, y^n)\ \varepsilon\, {}_2A_i^{*t}$$

and (6.4.4)

$$(y^1, \ldots, y^{t-1}, u_i^t, y^{t+1}, \ldots, y^n)\ \not\varepsilon\, {}_1A_i^t\,.$$

The latter could happen only if there exists a $j \neq i$, such that

$$\begin{aligned} &P\{v(u_j) = (y^1, \ldots, y^{t-1}, u_i^t, y^{t+1}, \ldots, y^n) \mid s_n^*\} \\ &\geqq P\{v(u_1) = (y^1, \ldots, y^{t-1}, u_i^t, y^{t+1}, \ldots, y^n) \mid s_n^*\} \end{aligned} \tag{6.4.5}$$

and consequently

$$(y^1, \ldots, y^{t-1}, u_i^t, y^{t+1}, \ldots y^n)\ \not\varepsilon \bigcup_{i=1}^{N} A_i\,.$$

In either case we can find a $j \neq i$ such that

$$\begin{aligned} &P\{v(u_j) = (y^1, \ldots, y^{t-1}, u_i^t, y^{t+1}, \ldots, y^n) \mid s_n^*\} \\ &\geqq P(v(u_i) = (y^1, \ldots, y^{t-1}, u_i^t, y^{t+1}, \ldots, y^n) \mid s_n^*\}\,. \end{aligned} \tag{6.4.6}$$

In case the points $1'$, and $2'$, coincide, the lemma holds. Assume therefore that $1'$ and $2'$ do not coincide. Hence $w'(1|2) \neq w'(1|1)$, and hence $w'(1|1) > 0$. For, if the latter were not true, the points $1'$, $2'$, and 2 of our diagram would coincide at the top of the ordinate axis.

Suppose first that, for the j in (6.4.6), $u_j^t = 1$. Multiplying both sides of (6.4.6) by $w'(2|1)\,[w'(1|1)]^{-1} \geqq 0$, we obtain that

$$\begin{aligned} &P\{v(u_j) = (y^1, \ldots, y^{t-1}, 2, y^{t+1}, \ldots, y^n) \mid s_n^*\} \\ &\geqq P\{v(u_i) = (y^1, \ldots, y^{t-1}, 2, y^{t+1}, \ldots, y^n) \mid s_n^*\}\,. \end{aligned} \tag{6.4.7}$$

This contradicts the fact that $(y^1, \ldots, y^{t-1}, y^{t+1}, \ldots, y^n)\ \varepsilon\, {}_2A_i^{*t}$ and proves (6.4.2) when $u_j^t = 1$.

Suppose now that, for the j in (6.4.6), $u_j^t = 2$. From the fact that $w'(1|2) \leqq w'(1|1)$, $w'(1|1) > 0$, and (6.4.6), we obtain that

$$\begin{aligned} &P\{v(u_j^1,\ldots,u_j^{t-1},u_j^{t+1},\ldots,u_j^n) = (y^1,\ldots,y^{t-1},y^{t+1},\ldots,y^n) \mid s_n^*\} \\ &\geqq P\{v(u_i^1,\ldots,u_i^{t-1},u_i^{t+1},\ldots,u_i^n) = (y^1,\ldots,y^{t-1},y^{t+1},\ldots,y^n) \mid s_n^*\}. \end{aligned} \tag{6.4.8}$$

However, $w'(2|2) \geqq w'(2|1)$ and (6.4.8) imply that

$$\begin{aligned} &P\{v(u_j) = (y^1,\ldots,y^{t-1},2,y^{t+1},\ldots,y^n) \mid s_n^*\} \\ &\geqq P\{v(u_i) = (y^1,\ldots,y^{t-1},2,y^{t+1},\ldots,y^n) \mid s_n^*\}. \end{aligned} \tag{6.4.9}$$

This also contradicts the fact that $(y^1,\ldots,y^{t-1},y^{t+1},\ldots y^n) \varepsilon\, {}_2A_i^{*t}$. This proves (6.4.2) when $u_j^t = 2$, and completes the proof when $u_i^t = 1$. The proof when $u_i^t = 2$ is (symmetrically) the same.

We now prove (6.4.3) inductively. Assume that it holds for $s_n' \varepsilon S_n$. We shall show that then (6.4.3) holds for s_n, where s_n is obtained from s_n' by changing the element w' in the kth component of s_n' to w.

Define $B(i,k)$ by

$${}_1A_i^{*k} = {}_2A_i^{*k} \cup B(i,k), \quad {}_2A_i^{*k} \cap B(i,k) = \phi.$$

This definition is possible because of (6.4.2). Let

$$\begin{aligned} P\{v(u_i^1,\ldots,u_i^{k-1},u_i^{k+1},\ldots,u_i^n) \varepsilon B(i,k) \mid s_n'\} &= a_1' \\ P\{v(u_i^1,\ldots,u_i^{k-1},u_i^{k+1},\ldots,u_i^n) \varepsilon\, {}_2A_i^{*k} \mid s_n'\} &= a_2'. \end{aligned}$$

If $u_i^k = 1$, then

$$P\{v(u_i) \varepsilon A_i \mid s_n'\} = w'(1|1)\, a_1' + a_2'.$$

This is true because, if $(y^1,\ldots,y^{k-1},y^{k+1},\ldots,y^n)$ is in ${}_2A_i^{*k}$, both $(y^1,\ldots,y^{k-1},1,y^{k+1},\ldots,y^n)$ and $(y^1,\ldots,y^{k-1},2,y^{k+1},\ldots,y^n)$ are in A_i. For the same reason, if $u_i^k = 2$, then

$$P\{v(u_i) \varepsilon A_i \mid s_n'\} = w'(2|2)\, a_1' + a_2'.$$

Now, if in the kth component of s_n' we replace w' by w, we get, in both cases, that

$$P\{v(u_i) \varepsilon A_i \mid s_n\} \geqq P\{v(u_i) \varepsilon A_i \mid s_n'\},$$

because $w(1|1) \geqq w'(1|1)$ and $w(2|2) \geqq w'(2|2)$. This completes the proof of the lemma.

For any c.p.f. w define

$$R(\pi, w) = H(\pi') - \sum_i \pi_i H(w(\cdot|i)),$$

where H, π, and π' are as in (2.2.1). We have already seen (just below (2.2.1)) that, for fixed w, $R(\pi,w)$ is a concave function of π. Let us now prove that, for fixed π, $R(\pi,w)$ is a convex function of w. Let w_1 and w_2

be two c.p.f.'s. Let Z be a chance variable which takes the values 1 and 2. The chance variable X is always to have the distribution π, independently of the value of Z. When $Z = i$ and $X = j$, the chance variable Y is to have the distribution $w_i(\cdot|j)$. Thus the statement that $R(\pi, w)$ is convex in w is equivalent to saying that

$$H(Y|Z) - H(Y|X, Z) \geqq H(Y) - H(Y|X).$$

The left member of this equals

$$H(X|Z) - H(X|Y, Z) = H(X) - H(X|Y, Z)$$

since Z and X are independent. The right member of the inequality to be proved equals $H(X) - H(X|Y)$. Since $H(X|Y, Z) \leqq H(X|Y)$, the desired result follows.

Lemma 6.4.2. We have

$$\max_{\pi} \min_{w \varepsilon \bar{S}} R(\pi, w) = \min_{w \varepsilon \bar{S}} \max_{\pi} R(\pi, w) = C'.$$

Proof: $\bar{S}$ and $\{\pi\}$ are normcompact convex sets, and $R(\pi, w)$ is normcontinuous in both variables. The first equality then follows from the minimax theorem of KAKUTANI [1]. The second is a consequence of Theorems 3.2.1 and 3.3.1.

Lemma 6.4.3. An (n, N, λ) code for the channel S_n is also an (n, N, λ) code for the channel $(\bar{S})_n$. S need not be finite. The converse is trivially true.

Proof: Any element of $\bar{S}$ can be approximated arbitrarily closely by a matrix, each of whose rows, say the jth, is a convex linear combination of the extreme points of $T(j)$. Any element of $(\bar{S})_n$ can be approximated arbitrarily closely by a convex linear combination of n-tuples of such matrices. The lemma follows.

We now prove the theorem in the case where $a = 2$. If $T(1)$ and $T(2)$ are not disjoint, obviously both C and C' are zero. We may therefore assume that $T(1)$ and $T(2)$ are disjoint. Let w' be the c.p.f. whose capacity is C'. A code for the arbitrarily varying channel is also a code for the d.m.c. with c.p.f. w'. Without loss of generality we may assume that w' has the representation

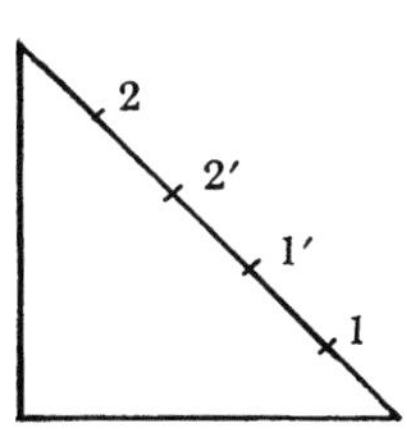

1′, 2′ in the above diagram. Let 1 (resp., 2) in this diagram be the other end point of $T(1)$ (resp., $T(2)$). From the application to the d.m.c. made immediately after the proof of Theorem 7.3.2 (it is more convenient to prove this theorem in greater generality in Chapter 7), we obtain that there exists a s.m.l.c.

$$(n, \exp_2\{nC' - \sqrt{n}K(\lambda)\}, \lambda)$$

for the d.m.c. with c.p.f. w'. From Lemma 6.4.1 we conclude that this is a code for the arbitrarily varying channel.

It now remains to prove the theorem for general a. For each pair (i, i') let $C(i, i')$ be the capacity of the matrix whose rows are the point of $T(i)$ nearest to $T(i')$ and the point of $T(i')$ nearest to $T(i)$. (When $T(i)$ and $T(i')$ are not disjoint these points may not be uniquely defined, but $C(i, i') = 0$ anyhow.) We will prove that

$$C = C' = \max_{(i, i')} C(i, i') . \tag{6.4.10}$$

Let M be an $(a \times b)$ stochastic matrix. Suppose all rows are convex linear combinations of two (extreme) rows. Let M' be the $(2 \times b)$ matrix of these rows. Then the capacity of M equals the capacity of M'. This follows from the fact that the capacity of M is $\max_{\pi} R(\pi, M)$ and the fact that $H(w)$ is a concave function of the c.p.f. w. For suppose π_0 is a distribution on the a rows. The resulting π_0' is the same as the π' which results from a suitable distribution π on the two extreme rows. On the other hand,

$$\sum_{i=1}^{a} \pi_{0i} H(w(\cdot|i)) \geq \Sigma \pi_i H(w(\cdot|i)) ,$$

where the second summation is over the extreme rows only, because of the concavity of H as a function of w.

We now want to define some very temporary but time saving terminology. Let us agree to say that, in the last diagram of this section, 1 is to the right of 1′, 2′, and 2, etc., and that 2′, for example, is to the left of 1 and 1′. Let z (resp., z') be the farthest to the right (resp., to the left) of the left (resp., right) ends of the intervals $T(i)$, $i = 1, \ldots, a$. Without loss of generality we assume that z is the left end of $T(1)$ and z' is the right end of $T(2)$.

If z is to the left of z', the second and third members of (6.4.10) are both zero, and the first is zero by Theorem 6.2.1. It remains, therefore, to consider only the case when z is to the right of z'.

Then the capacity of the matrix with two rows, z and z', is the third member of (6.4.10). By the paragraph which immediately follows

(6.4.10), it is also the second member. Using only the letters 1 and 2 we see that C, if it exists, must be at least C'. It remains to prove that $C = C'$. Suppose all the letters of the input alphabet, not just 1 and 2, were used in the words sent. If a letter other than 1 or 2 were sent, the jammer could so randomize the choice of c.p.f. to govern the transmission of this letter that it would be equivalent to sending a random letter with distribution on 1 and 2. Again we use the paragraph after (6.4.10) to conclude that $C = C'$. This completes the proof of the theorem.

An easy way to compute C is therefore the following: If z is to the left of z', $C = 0$. Otherwise C is the capacity of the matrix with rows z and z'.

6.5. Certain results for the general arbitrarily varying channel. In this section we will describe without proof two recent results on general arbitrarily varying channels (a, b, c finite), due, respectively, to Dobrushin and Stambler [1] and Stambler [1]. The proofs are very long and consist of very detailed computations of average probability of error along the lines of Theorem 7.3.1 below. The authors stand out for the boldness of their conjectures and their willingness to perform difficult calculations of such great length, when the truth of the initial conjecture is not known until the final computations are performed. We omit the proofs because a) they are very long and their inclusion would substantially increase the length of the present monograph, and b) we are unable to improve or abbreviate them appreciably.

We begin with the result of Dobrushin and Stambler [1]. The code requirement on the probability of error in previous sections was

$$\sup_{j} \sup_{s_n} P\{v(u_i) \notin A_i \mid s_n\} \leq \lambda \tag{6.5.1}$$

for $i = 1, \ldots, N$; $s_n \in \bar{S}_n$. The corresponding requirement of the work now being described is

$$\sup_{s_n} \frac{1}{N} \sum_{i=1}^{N} P\{v(u_i) \notin A_i \mid s_n\} \leqq \lambda \,, \tag{6.5.2}$$

i.e., average error rather than maximal error is bounded. Clearly, a code which satisfies (6.5.1) satisfies (6.5.2), but not necessarily conversely. Define $R(\pi, w)$ as in Section 6.4,

$$C_\pi(\bar{S}) = \min_{w \in \bar{S}} R(\pi, w) \,.$$

By Lemma 6.4.2,

$$C' = \max_{\pi} C_\pi(\bar{S}) \,.$$

We now need a large number of definitions. Define the probability distribution $\pi_s(\cdot)$ on Y by

$$\pi_s(j) = \sum_{i \in X} \pi(i) w(j \mid i \mid s) \,. \tag{6.5.3}$$

Let R^* be the totality of all c.p.f.'s $r(\cdot|\cdot,\cdot)$ from $X \times S$ to Y, so that, for any such r,

$$r(j|i, s) \geqq 0, \sum_{j=1}^{b} r(j|i, s) = 1 . \tag{6.5.4}$$

Let Q^* be the totality of all c.p.f.'s $q(\cdot|\cdot)$ from X to S, i.e., for any such q,

$$q(s|i) \geq 0, \sum_{s} q(s|i) = 1 . \tag{6.5.5}$$

For fixed $i \varepsilon X$ and $s \varepsilon S$ define, for $r \varepsilon R^*$,

$$\overline{H}_\pi(r(\cdot|i, s)|\pi_s(\cdot)) = \sum_{j \varepsilon Y} r(j|i, s) \log \frac{r(j|i, s)}{\pi_s(j)} . \tag{6.5.6}$$

For $r \varepsilon R^*$ and $q \varepsilon Q^*$ define

$$\overline{H}(\pi, q, r) = \sum_{i \varepsilon X} \sum_{s \varepsilon S} \pi(i) q(s|i) \overline{H}_n(r(\cdot|i, s)|\pi_s(\cdot)) . \tag{6.5.7}$$

It can be shown that $\overline{H}_\pi \geq O$, and hence $\overline{H} \geq O$. Denote by $V(w)$, $w \varepsilon \bar{S}$, the totality of all pairs (q,r), $q \varepsilon Q^*$, $r \varepsilon R^*$, such that, for all i and j,

$$w(j|i) = \sum_{s \varepsilon S} q(s|i) r(j|i,s) . \tag{6.5.8}$$

Note that, for any $w \varepsilon \bar{S}$, $V(w) \neq \emptyset$, since, for any $q \varepsilon Q^*$ and $r(j|i, s) \equiv w(j|i)$, the pair $(q, r) \varepsilon V(w)$. Finally define

$$H(\pi, S) = \inf_{w \varepsilon \bar{S}} \inf_{(q,r) \varepsilon V(w)} H(\pi, q, r) . \tag{6.5.9}$$

Theorem 6.5.1. If there exists a π such that

$$C_\pi(\bar{S}) = C' \tag{6.5.10}$$

and

$$H(\pi, S) > 0 , \tag{6.5.11}$$

then, for any $0 < \lambda < 1$ and $\varepsilon > 0$, there exists, for n sufficiently large, a code for the arbitrarily varying channel $(\bar{S})_n$ which has length $\exp_2\{n[C' - \varepsilon]\}$ and satisfies (6.5.2).

Theorem 6.5.2. For any $0 < \lambda < 1$ and $\varepsilon > 0$, there does not exist, when n is sufficiently large, a code for the arbitrarily varying channel $(\bar{S})_n$ which has length $\exp_2\{n[C' + \varepsilon]\}$ and satisfies (6.5.2).

Theorem 6.5.2 does not require (6.5.11) and, indeed, is quite obvious. This is so because the jammer can always use, for each letter, the channel in $\bar{S}$ whose capacity is a minimum. The desired result then follows from Theorem 3.3.1 and Lemma 3.1.1. It is Theorem 6.5.1 which is so difficult to prove.

Condition (6.5.11) is sufficient but not necessary for Theorem 6.5.1. An example is given in the paper cited to show this. An example is also given in which the condition is necessary, i.e., the capacity is zero, although $C' > 0$. Of course, $H(\pi, S) = 0$ for the maximizing π in this example.

The same authors also give a condition equivalent to (6.5.11), which may be useful in many instances. A consequence of this condition is the following: Let w_0 be a c.p.f. with positive capacity. If S consists of c.p.f.'s which lie in a sufficiently small neighborhood of w_0, the condition (6.5.11) is fulfilled.

We now proceed to a description of the results of STAMBLER [1]. By a *randomized* code is meant a code for which, when the sender wants to send the ith word, $i = 1, \ldots, N$, he chooses the sequence to be actually sent by a probability distribution q_i on X_n. For a randomized code the conditions (6.5.1) and (6.5.2) correspond, respectively, to

$$\sup_i \sup_{s_n} \sum_{x_n \varepsilon X_n} q_i(x_n) P\{v(x_n) \notin A_i | s_n\} \leqq \lambda , \tag{6.5.12}$$

and

$$\sup_{s_n} \frac{1}{N} \sum_{i=1}^{N} \sum_{x_n \varepsilon X_n} q_i(x_n) P\{v(x_n) \notin A_i | s_n\} \leqq \lambda . \tag{6.5.13}$$

We already know what it means to say that the sender does not know s_n and the receiver does. In that case A_i in (6.5.1), (6.5.2), (6.5.12), and (6.5.13) is to be replaced by $A_i(s_n)$, say.

The next three theorems (due to STAMBLER [1]) *all* postulate that the receiver, but not the sender, knows the s_n which governs the transmission of any word. We will call this condition R^+. We remind the reader of the definition of $\overline{C}$ in (4.1.3).

Theorem 6.5.3. Let $0 < \lambda < 1$ and $\varepsilon > 0$ be arbitrary. Under conditions R^+ and (6.5.2) there exists, for n sufficiently large, a code of length $\exp_2\{n[\overline{C} - \varepsilon]\}$ for the arbitrarily varying channel S_n.

Theorem 6.5.4. Let $0 < \lambda < 1$ and $\varepsilon > 0$ be arbitrary. Under conditions R^+ and (6.5.12) there exists, for n sufficiently large, a randomized code of length $\exp_2\{n[\overline{C} - \varepsilon]\}$ for the arbitrarily varying channel S_n.

Theorem 6.5.5. Under conditions R^+ and (6.5.13), the length N of any randomized code for the arbitrarily varying channel S_n, satisfies

$$\log N \leqq \frac{n\overline{C} + 1}{1 - \lambda} , \tag{6.5.14}$$

A result of this kind is called a weak converse; more in Chapter 7. The inequality (6.5.14) is the better the smaller is λ. Since a non-randomized code is a special case of a randomized code, the last theorem is stronger in this respect. Since the condition (6.5.13) is weaker than condition (6.5.12), the theorem is stronger than it would be if (6.5.12) were postulated.

Remarks

The theorems of Section 6.2 are due to KIEFER and WOLFOWITZ [1] and are the first results in the subject. The counter-example of Section 6.3 is due to AHLSWEDE and WOLFOWITZ [1]. Theorem 6.4.1 is due to AHLSWEDE and WOLFOWITZ [2]. Lemma 6.4.2 is due to STIGLITZ [1].

Theorems 6.5.1 and 6.5.2 are due to DOBRUSHIN and STAMBLER [1]. Theorems 6.5.3–6.5.5 are due to STAMBLER [1].

AHLSWEDE [1] has recently announced the following results: a) If the capacity $\bar{C}_1$ of an arbitrarily varying channel, with definition of error given by (6.5.2), is positive, it is equal to

$$\min_{\pi} \max R(\pi, w) ,$$

where the minimum is over the convex hull of S, which need not be finite. b) The following are always equal for any set S, which need not be finite: 1) $\bar{C}_1$ 2) the capacity of a randomized code, with definition of error either (6.5.1) or (6.5.2). If they are positive then they equal the expression in a) above.

7. General Discrete Channels

7.1. Alternative description of the general discrete channel. In the definition of the general discrete channel given in Section 5.1 we imposed no consistency conditions (except the obvious ones) on the totality $\{P_n\{v(u_0) = v_0\}\}$. Clearly, however, the definitions and notations were designed for the imposition of consistency conditions, and the remainder of Chapter 5 and all of Chapter 6 are based on various sets of consistency conditions. These consistency conditions were particularly concerned with the variation of n or, to put it differently, with the transmission and reception of individual letters and blocks of letters. When these questions do not explicitly arise the definitions and notation of Section 5.1 are unnecessarily involved and complicated. For example, if n does not vary why not simply consider the totality $\{P_n\{v(u_0) = v_0\}\}$ with n fixed? In this chapter we will do exactly that, and obtain results which can then be specialized to various channels. Our new definitions and notation will not only be simpler for our purpose but perhaps also more suggestive. The obvious correspondence with the definitions and notation of Section 5.1 will be indicated.

Let G_1 be a set of g_1 elements, and G_2 a set of g_2 elements. It will simplify the notation and not cause any loss of generality if we let G_1 be the set of integers $1, \ldots, g_1$, and let G_2 be the set of integers $1, \ldots, g_2$. Let $h(j \mid i)$ be a non-negative function (c.p.f.) defined for every pair (i, j) in $G = (G_1 \times G_2)$, and such that

$$\sum_{j=1}^{g_2} h(j \mid i) = 1, \quad i = 1, \ldots, g_1. \tag{7.1.1}$$

Call a code (N, λ), $0 \leq \lambda \leq 1$, a system

$$\{(u_1, A_1), \ldots, (u_N, A_N)\}, \tag{7.1.2}$$

where now $u_1, \ldots, u_N$ are elements of G_1, $A_1, \ldots, A_N$ are disjoint subsets of G_2, and

$$\sum_{j \varepsilon A_i} h(j \mid u_i) \geqq 1 - \lambda, \quad i = 1, \ldots, N. \tag{7.1.3}$$

It is easily seen that this description corresponds to the description of Section 5.1 if in the latter we fix n, and then make the following correspondences: g_1 as the number of n-sequences in D'_n, g_2 as the number of sequences in D''_n, and h as the probability distribution $P_n\{\ \}$.

Let Q' be any probability distribution (stochastic input) on G_1. Define the probability distribution Q (joint distribution) on G by

$$Q(i, j) = Q'(i)\, h(j \mid i), \quad (i, j) \varepsilon G \tag{7.1.4}$$

and the probability distribution Q'' (stochastic output) on G_2 by

$$Q''(j) = \sum_{i=1}^{g_1} Q(i, j). \tag{7.1.5}$$

Let $Q'(\cdot \mid \cdot)$ and $Q''(\cdot \mid \cdot)$ be defined in complete analogy with $Q'_n(\cdot \mid \cdot)$ and $Q''_n(\cdot \mid \cdot)$ of Chapter 5. Define the function I on G exactly as in Section 5.1, i.e., if $Q(i, j) = 0$ then $I(i, j) = 0$, and otherwise

$$I(i, j) = \log \frac{Q(i, j)}{Q'(i)\, Q''(j)}. \tag{7.1.6}$$

When we wish to put Q' into evidence we write $I(Q')$ and $I(i, j \mid Q')$. Let (u, v) be a chance variable with values in G and distribution Q. Let J be the chance variable $I(u, v)$. If we wish to put Q' into evidence we write $J(Q')$. As in Chapter 5, we call $R(Q') = E(J(Q'))$ the transmission rate of the channel for the stochastic input Q'.

7.2. The method of maximal codes. We now prove several coding theorems by different variations of the same basic method. Essential in this method is the use of codes of maximal length. We have already used this idea in Chapters 3 and 4.

Theorem 7.2.1. Let θ and $\lambda < 1$ be any positive numbers. Let Q' be any distribution (stochastic input) on G_1. Let $W \subset G$ be the set defined by

$$W = \{(i, j) \mid I(i, j) > \theta\}$$

and let $\overline{W}$ be its complement. There exists an (N, λ) code for the channel such that

$$N > 2^{\theta} [\lambda - Q(\overline{W})]. \tag{7.2.1}$$

(The case $\lambda = 1$ is of course trivial. There is always a code $(g_1, 1)$.)

Proof: We may assume that

$$Q(W) > 1 - \lambda \tag{7.2.2}$$

or the theorem is trivially true. For any $i \varepsilon G_1$ define the subset $B(i)$ of G_2 by

$$B(i) = \{j \mid (i, j) \varepsilon W\}.$$

We now prove that there is an element $u_1 \varepsilon G_1$ such that

$$P\{v(u_1) \varepsilon B(u_1)\} = \sum_{j \varepsilon B(u_1)} h(j \mid u_1) \geqq 1 - \lambda. \tag{7.2.3}$$

Suppose this were not so. Then, for every $i \varepsilon G_1$ we would have

$$P\{v(i) \varepsilon B(i)\} < 1 - \lambda. \tag{7.2.4}$$

From (7.2.4) we at once obtain $Q(W) < 1 - \lambda$, which contradicts (7.2.2) and proves the existence of a u_1 which satisfies (7.2.3).

Now let

$$\{(u_1, A_1), \ldots, (u_N, A_N)\} \tag{7.2.5}$$

be a code (N, λ) such that:

$$A_j = B(u_j) - (A_1 \cup \cdots \cup A_{j-1}), \quad j = 1, \ldots, N \tag{7.2.6}$$

it is impossible to add an element (u_{N+1}, A_{N+1}) to the code (7.2.5) such that the resulting set is a code $(N+1, \lambda)$ and (7.2.7)

$$A_{N+1} = B(u_{N+1}) - (A_1 \cup \cdots \cup A_N).$$

Then, for any $i \varepsilon G_1$, we have

$$P\{v(i) \varepsilon (B(i) - (A_1 \cup \cdots \cup A_N))\} < 1 - \lambda. \tag{7.2.8}$$

From (7.2.8) we obtain

$$Q\{W - [A_1 \cup \cdots \cup A_N]\} < 1 - \lambda, \tag{7.2.9}$$

where $[A_1 \cup \cdots \cup A_N]$ is the cylinder set in G with base $A_1 \cup \cdots \cup A_N$.

From the definition of W we have, for $j = 1, \ldots, N$ and $v_0 \varepsilon B(u_j)$,

$$P\{v(u_j) = v_0\} = h(v_0 \mid u_j) > 2^{\theta} \cdot Q''(v_0). \tag{7.2.10}$$

Summing both sides of (7.2.10) over all v_0 in $B(u_j)$ we obtain

$$P\{v(u_j) \varepsilon B(u_j)\} = Q''\{B(u_j) \mid u_j\} > 2^{\theta} \cdot Q''(B(u_j)). \tag{7.2.11}$$

The left member of (7.2.11) is not greater than one. Hence

$$Q''(B(u_j)) < 2^{-\theta}. \tag{7.2.12}$$

From (7.2.9) and (7.2.12) we obtain

$$Q(W) < (1 - \lambda) + N \cdot 2^{-\theta},$$

which is exactly (7.2.1). This proves Theorem 7.2.1.

The simplicity of application of Theorem 7.2.1 is due to the fact that the channel enters only through the distribution of $J(Q')$. Consider the d.m.c. with notation as in Chapter 3. Let $u = (X_1, \ldots, X_n)$, and $X_1, \ldots, X_n$ be independent, identically distributed chance variables with distribution $\bar{\pi}$. Then J is the sum of n independent chance variables, each with mean C and variance σ^2, say. Let $\theta = nC - d\sigma\sqrt{n}$, where $d > 0$ is such that, for every n,

$$P\{J \leqq nC - d\sigma\sqrt{n}\} < \lambda' < \lambda. \qquad (7.2.13)$$

Such a number d exists, by CHEBYSHEV's inequality. We obtain from Theorem 7.2.1 that, for every n, there exists a code $(n, 2^{nC-K\sqrt{n}}, \lambda)$ for the d.m.c., where $K > 0$ is a suitable constant. We can even do slightly better if $\lambda > \frac{1}{2}$. We then choose $\lambda' > \frac{1}{2}$. It follows from the central limit theorem that, when n is sufficiently large, (7.2.13) is satisfied for a negative value of d. Thus, when $\lambda > \frac{1}{2}$ and n is sufficiently large, there exists a code $(n, 2^{nC+K'\sqrt{n}}, \lambda)$ for the d.m.c., where $K' > 0$ is a suitable constant.

It is interesting to obtain a bound on σ of the preceding paragraph when each of the X's has the distribution π (not necessarily $\bar{\pi}$, therefore). Obviously then

$$\begin{aligned}\sigma^2 &\leqq \sum_{i,j} \pi_i w(j \mid i) \left[\log\left(\frac{\pi_i w(j \mid i)}{\pi_i \pi'_j}\right)\right]^2 \\ &= \sum_j \pi'_j \left\{\sum_i \frac{\pi_i w(j \mid i)}{\pi'_j} \left[\log\left(\frac{\pi_i w(j \mid i)}{\pi'_j}\right)\right]^2\right\} \\ &\quad + \sum_i \pi_i \log^2 \pi_i - 2\sum_{i,j} \pi_i w(j \mid i) \left[\log \frac{\pi_i w(j \mid i)}{\pi'_j}\right] \log \pi_i .\end{aligned}$$

An argument with a Lagrange multiplier shows that the second term on the right is $\leqq (\log a)^2$ when $a \geqq 3$, and $\leqq 2.343$ when $a = 2$. Hence also, for every j, the quantity in curly brackets in the first term on the right is also bounded by these bounds, and therefore so is the entire first term. The third term is obviously non-positive. Thus

$$\sigma^2 \leqq 2(\log a)^2 \qquad (7.2.14)$$

when $a \geqq 3$, and

$$\sigma^2 \leqq 4.686$$

when $a = 2$. Consequently, if a is large, the code length obtained through the use of (7.2.13) and Theorem 7.2.1 is larger than the code length obtained in Theorem 3.2.1. (If the input and output alphabets have different numbers of letters, let c_0 be the smaller of these two numbers. The above argument then shows (because of the symmetry of the first inequality for σ^2 in i and j) that

$$\sigma^2 \leqq 2(\log c_0)^2$$

when $c_0 \geqq 3$, and

$$\sigma^2 \leqq 4.686$$

when $c_0 = 2$.)

There is an extension of Theorem 7.2.1 which is trivial to obtain but which is useful in applications. Let L_0 be a subset of G_1, and let $L_{00} = (L_0 \times G_2)$. If, in the statement of Theorem 7.2.1, we replace the set $\overline{W}$ by the set $\overline{W} \cup L_{00}$, the resulting theorem is trivially true, and, indeed, is weaker than Theorem 7.2.1. However, if we examine the proof of Theorem 7.2.1 with $\overline{W}$ replaced by $\overline{W} \cup L_{00}$, we can conclude that the code (7.2.5), whose existence is asserted by the theorem, is such that no u_i is a member of L_0. This remark, to which, for ease of citation, we shall refer as the *corollary to Theorem 7.2.1*, has the following application which will be illustrated in a later chapter: Suppose the code elements $u_1, \ldots, u_N$ are to be members of G_1, but a distribution Q' on a set $G_0 \supset G_1$ is convenient to use. By setting $L_0 = G_0 - G_1$ we can employ the distribution Q' while effectively restricting the u_i to be members of G_1.

We shall now consider the compound channel with t c.p.f.'s, $h(\cdot \mid \cdot \mid 1), \ldots, h(\cdot \mid \cdot \mid t)$, and c.p.f. unknown to both sender and receiver. Let Q' be any distribution (stochastic input) on G_1. Denote Q, Q'', I, and J which correspond to $h(\cdot \mid \cdot \mid s)$ by the superscript s, $s = 1, \ldots, t$. We will now prove

Theorem 7.2.2. Let θ, θ', and $\lambda < 1$ be arbitrary positive numbers. Under the conditions and with the notation of the preceding paragraph there exists an (N, λ) code for the compound channel such that

$$N > \frac{2^\theta}{t}\left[\lambda - t^2 \cdot 2^{-\theta'} - \sum_{s=1}^{t} Q^{(s)}\{(i,j) \mid I(i,j) \leqq \theta + \theta'\}\right]. \quad (7.2.15)$$

The idea which underlies the proof of this theorem is that a code $\left(N, \frac{\lambda}{t}\right)$ for the single c.p.f.

$$h^*(j \mid i) = \frac{1}{t} \sum_{s=1}^{t} h(j \mid i \mid s), \; i = 1, \ldots, g_1; \; j = 1, \ldots, g_2 \quad (7.2.16)$$

is a code (N, λ) for the compound channel. Let V be the set defined by

$$V = \left\{(i,j) \,\middle|\, \sum_{s=1}^{t} h(j \mid i \mid s) \leqq 2^\theta \sum_{s=1}^{t} Q''^{(s)}(j)\right\}.$$

Theorem 7.2.2 will follow at once from the next two lemmas.

Lemma 7.2.1. Under the hypotheses of Theorem 7.2.2 there exists a code (N, λ) for the compound channel such that

$$N > \frac{2^\theta}{t}\left[\lambda - \sum_{s=1}^{t} Q^{(s)}(V)\right]. \quad (7.2.17)$$

This lemma follows immediately from the application of Theorem 7.2.1 to the c.p.f. (7.2.15).

Lemma 7.2.2. We have

$$\sum_{s=1}^{t} Q^{(s)}(V) \leqq t^2 \cdot 2^{-\theta'} + \sum_{s=1}^{t} Q^{(s)}\{(i,j) \mid I^{(s)}(i,j) < \theta + \theta'\}. \quad (7.2.18)$$

Proof: Consider the set

$$V'_s = \{(i,j) \mid I^{(s)}(i,j) \geqq \theta + \theta'\} \cap V, \quad s = 1, \ldots, t. \quad (7.2.19)$$

For any (i, j) in V'_s we have

$$2^{\theta} \cdot \sum_{s=1}^{t} Q''^{(s)}(j) \geqq \sum_{s=1}^{t} h(j \mid i \mid s) \geqq h(j \mid i \mid s) \geqq 2^{\theta+\theta'} \cdot Q''^{(s)}(j). \quad (7.2.20)$$

Hence

$$Q''^{(s)}(j) \leqq 2^{-\theta'} \sum_{s=1}^{t} Q''^{(s)}(j). \quad (7.2.21)$$

Summing (7.2.20) over all j for which there exists an i such that (i, j) is in V'_s we obtain

$$Q^{(s)}(V'_s) \leqq t \cdot 2^{-\theta'}. \quad (7.2.22)$$

Now, for $s = 1, \ldots, t$,

$$Q^{(s)}(V) \leqq Q^{(s)}(V'_s) + Q^{(s)}\{(i,j) \mid I^{(s)}(i,j) < \theta + \theta'\}. \quad (7.2.23)$$

The lemma now follows from (7.2.21) and (7.2.22).

It is easy to apply Theorem 7.2.2 to a compound channel which consists of t d.m.c.'s and has capacity C. One application is as in Theorem 7.2.1, with $\theta = nC - 2d\sqrt{n}$ and $\theta' = d\sqrt{n}$, and d a suitable constant. If $\lambda > \frac{1}{2}$ and n is sufficiently large one can, as in the application of Theorem 7.2.1, obtain a positive multiple of $\sqrt{n}$ in the exponent of the expression for the length of the code. If a compound channel contains infinitely many c.p.f.'s one can let t be the number of c.p.f.'s in the canonical channel for any particular n. Thus the compound d.m.c. which contains infinitely many c.p.f.'s can also be treated by Theorem 7.2.2.

Now consider the compound channel of Theorem 7.2.2, but with the c.p.f. known to the sender (but not to the receiver). Let $Q'^{(1)}, \ldots, Q'^{(t)}$ be t distributions (stochastic inputs) on G_1. Define

$$Q^{(s)}(i,j) = Q'^{(s)}(i) \cdot h(j \mid i \mid s), \quad s = 1, \ldots, t; \quad i = 1, \ldots, g_1; \quad j = 1, \ldots, g_2$$

$$Q''^{(s)}(j) = \sum_{i=1}^{g_1} Q^{(s)}(i,j).$$

(Note the difference in definition between, for example, the present $Q^{(s)}$ and the $Q^{(s)}$ of Theorem 7.2.2. The stochastic input now depends on s.) We now write $I(Q'^{(s)})$ to put $Q'^{(s)}$ in evidence, and prove

Theorem 7.2.3. Let θ, θ', and $\lambda < 1$ be arbitrary positive numbers. Under the conditions and with the notation of the preceding paragraph there exists an (N, λ) code such that

$$N > \frac{2^{\theta}}{t}\left[\lambda - t^2 \cdot 2^{-\theta'} - \sum_{s=1}^{t} Q^{(s)}\{(i,j) \mid I(i,j \mid Q'^{(s)}) \leqq \theta + \theta'\}\right]. \quad (7.2.24)$$

Proof: Define G_1^* as the Cartesian product of G_1 with itself t times. Define the c.p.f. $\bar{h}$ on $G_1^* \times G_2$, as follows: Let $\bar{u}_0 = (u_0^{(1)}, \ldots, u_0^{(t)})$ be any element of G_1^*, and v_0 be any element of G_2. Then

$$\bar{h}(v_0 \mid \bar{u}_0) = \frac{1}{t} \sum_{s=1}^{t} h(v_0 \mid u_0^{(s)} \mid s).$$

It follows that a code $\left(N, \frac{\lambda}{t}\right)$ for the single c.p.f. $\bar{h}$ is a code (N, λ) for the compound channel of the present theorem. We now apply Theorem 7.2.1 to the single c.p.f. $\bar{h}$, with λ replaced by $\frac{\lambda}{t}$ and Q' replaced by $\bar{Q}' = Q'^{(1)} \times \ldots \times Q'^{(t)}$. We have

$$\bar{Q}(\bar{u}_0, v_0) = \frac{\bar{Q}'(\bar{u}_0)}{t} \sum_{s=1}^{t} h(v_0 \mid u_0^{(s)} \mid s) \quad (7.2.25)$$

$$\bar{Q}''(v_0) = \sum_{\bar{u}_0} \bar{Q}(\bar{u}_0, v_0) = \frac{1}{t} \sum_{s=1}^{t} Q''^{(s)}(v_0). \quad (7.2.26)$$

Hence

$$I(\bar{u}_0, v_0 \mid \bar{Q}') = \log \frac{\sum_{s=1}^{t} h(v_0 \mid u_0^{(s)} \mid s)}{\sum_{s=1}^{t} Q''^{(s)}(v_0)}. \quad (7.2.27)$$

Consequently the analogues of Lemmas 7.2.1 and 7.2.2 follow at once, and the theorem is then an immediate consequence.

The remarks made after Theorems 7.2.1 and 7.2.2 apply to Theorem 7.2.3 as well.

7.3. The method of random codes. We now describe another and, in practice, often more useful method of obtaining essentially the same result as Theorem 7.2.1. In the next paragraph we shall give a short summary of the method, and in the third paragraph we shall begin de novo and give a complete proof.

The method takes its name from the fact that one deals with an ensemble K_0 of codes, and there is a probability distribution on K_0 which assigns probabilities to the different codes of K_0. Each code in K_0 has the same length $2N$ but need not have probability of error $\leqq \lambda$. For any code $K \varepsilon K_0$, let $\bar{\lambda}(K)$ denote the average probability of error (see (3.1.9)).

Suppose that

$$E\bar{\lambda}(K) \leqq \frac{\lambda}{2}.$$

(Of course the operator E is with respect to the distribution on K_0.) Then there is obviously a code K' (say) in K_0 such that $\bar{\lambda}(K') \leqq \frac{\lambda}{2}$. The application of Lemma 3.1.1 to K' gives the existence of a code of length N and probability of error $\leqq \lambda$.

We now give the precise statements and proofs. We employ the channel described in Section 7.1 and all the notation described in that section. Let $(u_1^*, v(u_1^*)), \ldots, (u_N^*, v(u_N^*))$ be independent, identically distributed chance variables, with range in G and common probability distribution Q. We define $A_1^*, \ldots, A_N^*$, disjoint chance subsets of G_2, as follows: For $i = 1, \ldots, N$,

$$A_i^* = \{j \,\varepsilon\, G_2 \mid h(j \mid u_i^*) > h(j \mid u_k^*),\ k \neq i\}. \tag{7.3.1}$$

Consequently

$$A_i^* = \{j \varepsilon G_2 \mid I(u_i^*, j) > I(u_k^*, j),\ k \neq i\}. \tag{7.3.2}$$

Let $\bar{A}_i^*$ be the complement of A_i^* (in G_2). We now prove

Theorem 7.3.1. Let $\alpha > 1$ be arbitrary. We have

$$\frac{1}{N} E\left\{\sum_{i=1}^{N} \sum_{j \varepsilon \bar{A}_i^*} h(j \mid u_i^*)\right\} \leqq \frac{1}{\alpha} + Q\{(i, j) \mid I(i, j) \leqq \log \alpha N\}. \tag{7.3.3}$$

(The restriction $\alpha > 1$ is imposed only to avoid triviality; (7.3.3) is trivial for $0 \leqq \alpha \leqq 1$.)

The left member of (7.3.3) obviously equals

$$E\left\{\sum_{j \varepsilon \bar{A}_1^*} h(j \mid u_1^*)\right\}, \tag{7.3.4}$$

which it suffices henceforth to consider. Suppose $u_1^* = i$. In order that $j \varepsilon A_1^*$ we must have

$$I(i, j) > I(u_k^*, j),\ k \neq 1. \tag{7.3.5}$$

The chance variables $u_2^*, \ldots, u_N^*$ are distributed independently of each other and of u_1^*. Hence the probability of the relation (7.3.5) (for fixed i and j), is the $(N-1)^{\text{st}}$ power of the probability of the relation

$$I(i, j) > I(u_2^*, j). \tag{7.3.6}$$

The probability of (7.3.6) is $1 - \beta(i, j)$, where

$$\beta(i, j) = Q'\{i' \varepsilon G_1 \mid I(i, j) \leqq I(i', j)\}.$$

Consequently the quantity in (7.3.4) is equal to

$$\sum_{i,j} Q(i, j)\,[1 - (1 - \beta(i, j))^{N-1}]. \tag{7.3.7}$$

Define

$$B = \{(i, j) \mid I(i, j) > \log \alpha N\},$$

and $\bar{B}$ as the complement of B. Recall that, for $0 < x < 1$, $(1-x)^{N-1} > 1 - Nx$. We have

$$E\left\{\sum_{j \varepsilon \bar{A}_1^*} h(j \mid u_1^*)\right\} \leqq Q\{\bar{B}\} + N \cdot \sum_{(i,j)\varepsilon B} Q(i, j)\beta(i, j). \tag{7.3.8}$$

We now bound $\beta(i, j)$ for $(i, j) \varepsilon B$. If

$$I(i', j) \geqq I(i, j) > \log \alpha N, \tag{7.3.9}$$

then

$$\frac{Q(i', j)}{Q''(j)} > \alpha N Q'(i'). \tag{7.3.10}$$

Sum both sides of (7.3.10) over all i' in G_1 which satisfy (7.3.9). The left member of the sum is at most unity. The right member is $\alpha N \beta(i, j)$. Hence

$$\beta(i, j) < \frac{1}{\alpha N} \tag{7.3.11}$$

for $(i, j) \varepsilon B$. The desired result (7.3.3) follows from (7.3.8) and (7.3.11). This proves Theorem 7.3.1.

The following coding theorem is an almost immediate consequence of Theorem 7.3.1:

Theorem 7.3.2. If α and $N' = 2N$ are such that

$$\frac{1}{\alpha} + Q\{(i, j) \mid I(i, j) \leqq \log \alpha N'\} \leqq \frac{\lambda}{2} \leqq \frac{1}{2}, \tag{7.3.12}$$

then there exists a code (N, λ) as defined in (7.1.2).

Proof: It follows from (7.3.12) and Theorem 7.3.1 that there exists a set $(u_1^0, \ldots, u_{N'}^0)$ of values of $(u_1^*, \ldots, u_{N'}^*)$, and the corresponding set $(A_1^0, \ldots, A_{N'}^0)$ of sets $(A_1^*, \ldots, A_{N'}^*)$ such that

$$\frac{1}{N'} \sum_{i=1}^{N'} \sum_{j \notin A_i^0} h(j \mid u_i^0) \leqq \frac{\lambda}{2}. \tag{7.3.13}$$

Theorem 7.3.2 now follows from (7.3.13) and Lemma 3.1.1.

One can apply Theorem 7.3.2 to the discrete memoryless channel as follows: Fix n. Let G_1 be D'_n and G_2 be D''_n. Replace the function h by the appropriate value of the right member of (3.1.3). Let d and d' be such that

$$2^{-d\sqrt{n}+1} + 2P\left\{J \leqq nC - d'\sqrt{n}\right\} \leqq \lambda, \tag{7.3.14}$$

where the distribution of J is of course determined by Q'. Set $\alpha = 2^{d\sqrt{n}}$, $\alpha N' = 2^{nC - d'\sqrt{n}}$. Then Theorem 7.3.2 asserts the existence of a code $(2^{nC-(d+d')\sqrt{n}-1}, \lambda)$. Again, as in Section 7.2, if $\lambda > \frac{1}{2}$ we may take $(d + d')$ to be negative when n is sufficiently large.

We now give an almost trivial generalization of Theorem 7. 3. 1. This new form is what is needed for application to the theory of group codes for a binary symmetric channel (see Chapter 11) and to other subjects.

Theorem 7. 3. 3. The conclusion of Theorem 7. 3. 1 holds even if $(u_1^*, v(u_1^*)), \cdots, (u_N^*, v(u_N^*))$ are merely pairwise independent.

Proof: Inequality (7.3.8) holds even if $(u_1^*, v(u_1^*)), \cdots, (u_N^*, v(u_N^*))$ are merely pairwise independent, because the probability of a union of sets is not greater than the sum of the probabilities.

7.4. Weak converses. Throughout this section we assume that code (7.1.2), whose length we shall now bound above, is such that $N \geqq 2$ and $A_1 \cup \cdots \cup A_N = G_2$. The first of these assumptions is made to avoid the trivial. The second involves no loss of generality, because one can always add $G_2 - (A_1 \cup \cdots \cup A_N)$ to A_1, say. Define $\lambda(u_i)$ as in Chapter 3 by

$$P\{v(u_i) \varepsilon A_i\} = \sum_{j \varepsilon A_i} h(j \mid u_i) = 1 - \lambda(u_i), \; i = 1, \ldots, N,$$

and

$$\bar{\lambda} = \frac{1}{N} \sum_{i=1}^{N} \lambda(u_i).$$

We now prove

Theorem 7.4.1. Let Q_0' be the distribution on G_1 defined by

$$Q_0'(u_i) = \frac{1}{N}, \; i = 1, \ldots, N.$$

Then

$$H(Q_0' \mid Q_0'') \leqq \bar{\lambda} \log (N-1) - \bar{\lambda} \log \bar{\lambda} - (1-\bar{\lambda}) \log (1-\bar{\lambda}). \quad (7.4.1)$$

Proof: Let (u, v) be the chance variable (as in Section 7.1) with values in G and distribution Q_0. Let v' be the function of v defined as follows: When $v \varepsilon A_i$, $i = 1, \ldots, N$, $v' = i$. Then

$$H(Q_0' \mid Q_0'') = H(u \mid v) \leqq H(u \mid v') \quad (7.4.2)$$

by (2.2.16). Now, for $j = 1, \ldots, N$,

$$\begin{aligned} H(u \mid v' = j) \leqq & - P\{u = u_j \mid v' = j\} \log P\{u = u_j \mid v' = j\} \\ & - P\{u \neq u_j \mid v' = j\} \log P\{u \neq u_j \mid v' = j\} \quad (7.4.3) \\ & + P\{u \neq u_j \mid v' = j\} \log (N-1). \end{aligned}$$

Multiply both members of (7.4.3) by $P\{v' = j\}$ and sum with respect to j. For the last term in the right member we then obtain, temporarily neglecting the factor $\log (N-1)$,

$$\sum_j P\{v' = j\} \cdot P\{u \neq u_j \mid v' = j\} = 1 - \sum_j P\{u = u_j, v' = j\} = \bar{\lambda}.$$

Let $Z(u, v')$ be a function of u and v' defined as follows (with probability one): When $u = u_i$, $v' = i$, $i = 1, \ldots, N$, then $Z(u, v') = 1$; otherwise $Z(u, v') = 0$. Then

$$P\{Z = 0\} = \bar{\lambda}.$$

Now

$$\begin{aligned} &-\sum_j P\{u = u_j, v' = j\} \log P\{u = u_j \mid v' = j\} \\ &-\sum_j P\{u \neq u_j, v' = j\} \log P\{u \neq u_j \mid v' = j\} \\ &= -\sum_j P\{Z = 1, v' = j\} \log P\{Z = 1 \mid v' = j\} \\ &-\sum_j P\{Z = 0, v' = j\} \log P\{Z = 0 \mid v' = j\} \\ &= H(Z \mid v') \leqq H(Z) \\ &= -\bar{\lambda} \log \bar{\lambda} - (1 - \bar{\lambda}) \log (1 - \bar{\lambda}). \end{aligned}$$

Consequently

$$H(u \mid v') \leqq \bar{\lambda} \log (N - 1) - \bar{\lambda} \log \bar{\lambda} - (1 - \bar{\lambda}) \log (1 - \bar{\lambda})$$

which proves the theorem.

Suppose $g_1 = N = g_2$, and

$$h(j \mid i) = \frac{\bar{\lambda}}{N - 1}, \quad i \neq j$$

$$h(i \mid i) = 1 - \bar{\lambda}, \quad i = 1, \ldots, N.$$

Then $H(Q_0' \mid Q_0'')$ is equal to the right member of (7.4.1).

It is not necessary that the chance variable u have the uniform distribution Q_0' on $\{u_1, \ldots, u_N\}$; it could have any distribution. In that case $\bar{\lambda}$ should be replaced by $E[\lambda(u)]$.

Now let

$$T = \max R(Q'),$$

where the maximum, easily seen to be attained, is over all stochastic inputs Q'. Thus T is a function of the channel. We now prove

Theorem 7.4.2. For any channel a code (N, λ) satisfies

$$\log N \leqq \frac{T + 1}{1 - \bar{\lambda}} \leqq \frac{T + 1}{1 - \lambda} \tag{7.4.4}$$

Proof: Let Q_0' and $\bar{\lambda}$ be as used earlier in this section. Let $\lambda_M = \max_i \lambda(u_i)$. Then obviously $\lambda \geqq \lambda_M \geqq \bar{\lambda}$. From Theorem 7.4.1 we have

$$\begin{aligned} T \geqq R(Q_0') \geqq \log N + \bar{\lambda} \log \bar{\lambda} + (1 - \bar{\lambda}) \log (1 - \bar{\lambda}) \\ - \bar{\lambda} \log (N - 1). \end{aligned} \tag{7.4.5}$$

Since $-\bar{\lambda} \log \bar{\lambda} - (1 - \bar{\lambda}) \log (1 - \bar{\lambda}) \leqq 1$ we have at once that

$$T \geqq (1 - \bar{\lambda}) \log N - 1 \tag{7.4.6}$$

which is the desired result.

Suppose $G_1 = D'_n$, $G_2 = D''_n$. To emphasize the dependence on n we now write $T(n)$. Let ε and Δ be arbitrary positive numbers. Let

$$\lambda_n < \frac{n\varepsilon}{n\varepsilon + T(n)} - \Delta. \tag{7.4.7}$$

Then for all n sufficiently large there does not exist a code $(n, 2^{T(n)+n\varepsilon}, \lambda_n)$ for the channel. This result of course does not depend on the c.p.f. of the channel except through $T(n)$.

Since always $T(n) \leqq n \log a$, we can, for example, always assert the following: Let $\varepsilon > 0$ be arbitrary. Then for n sufficiently large there does not exist a code

$$\left(n, 2^{T(n)+n\varepsilon}, \frac{\varepsilon}{2(\varepsilon + \log a)}\right). \tag{7.4.8}$$

Theorem 7.4.2 is called the "weak converse" in distinction to the strong converses we have already proved for several channels.

Suppose now given a compound channel with a not necessarily finite set of c.p.f.'s. Each c.p.f. has a distinguishing index; let S be the totality of indices. We shall use the usual functions of a c.p.f. with the index of the c.p.f. appearing as a superscript, thus: $T^{(s)}$, $R^{(s)}(Q')$, etc. Suppose (7.1.2) is a code (N, λ) for the compound channel where neither sender nor receiver knows the c.p.f. being used for transmission. From (7.4.7) we have

$$R^{(s)}(Q'_0) \geqq (1-\lambda)\log N - 1. \tag{7.4.9}$$

Hence

$$\inf_s R^{(s)}(Q'_0) \geqq (1-\lambda)\log N - 1 \tag{7.4.10}$$

$$\max_{Q'} \inf_s R^{(s)}(Q') \geqq (1-\lambda)\log N - 1. \tag{7.4.11}$$

Call the left member of (7.4.11) $T_1(S)$. Thus we have proved

Theorem 7.4.3. Any code (N, λ) for the compound channel S where neither sender nor receiver knows the c.p.f. which governs the transmission of a word satisfies

$$\log N \leqq \frac{T_1(S)+1}{1-\bar{\lambda}} \leqq \frac{T_1(S)+1}{1-\lambda}. \tag{7.4.12}$$

Now consider the channel with totality S of c.p.f.'s where the receiver (but not the sender) knows the c.p.f. which governs the transmission of any word. The above argument, beginning with (7.4.9), holds in toto. Thus we have

Theorem 7.4.4. Any code (N, λ) for the compound channel S, where the receiver (but not the sender) knows the c.p.f. which governs the transmission of any word, satisfies (7.4.12).

Finally we consider the channel with totality S of c.p.f.'s, where the sender (but not the receiver) knows the c.p.f. which governs the transmission of any word. Define

$$T_2(S) = \inf_s \max_{Q'} R^{(s)}(Q').$$

Obviously $T_2(S) \geqq T_1(S)$. By an obvious proof like that of Theorem 4.5.4 we at once obtain from Theorem 7.4.2 the following

Theorem 7.4.5. Any code (N, λ) for the compound channel S, where the sender (but not the receiver) knows the c.p.f. which governs the transmission of any word, satisfies

$$\log N \leqq \frac{T_2(S)+1}{1-\bar{\lambda}} \leqq \frac{T_2(S)+1}{1-\lambda}. \tag{7.4.13}$$

7.5. Digression on the d.m.c. Consider the d.m.c. with notation as in Chapter 3. Let $R_n(Q'_n)$ be the quantity defined in Section 5.1. Our first object in this section will be to prove that

$$\max_{Q_n} R_n(Q'_n) = n\,C,$$

where C is, of course, the quantity defined in (3.1.11). This will be a consequence of Theorem 7.5.1 below.

Let $(X_1, Y_1), \ldots, (X_n, Y_n)$ be (not necessarily independent) vector chance variables each of whose components takes the values $1, \ldots, a$. Suppose

$$P\{Y_i = j \mid X_1, \ldots, X_n, Y_1, \ldots, Y_{i-1}\} = w(j \mid X_i), \quad i = 1, \ldots, n. \tag{7.5.1}$$

This condition means that, when $X = (X_1, \ldots, X_n)$ is fixed (at any n-sequence, of course), the Y_i are conditionally independently distributed, with distributions given by the right member of (7.5.1). This occurs in a d.m.c., for example. Write $Y = (Y_1, \ldots, Y_n)$. Define the marginal distributions

$$\left.\begin{array}{l} g_i(j) = P\{X_i = j\} \\ h_i(j) = P\{Y_i = j\} \end{array}\right. \quad i = 1, \ldots, n; \; j = 1, \ldots, a.$$

It follows from (7.5.1) that the distribution $g_i(\cdot)$ uniquely determines the distribution $h_i(\cdot)$.

Theorem 7.5.1. Under the above conditions

$$H(Y) - H(Y \mid X) \leqq n\,C. \tag{7.5.2}$$

The equality sign holds if and only if $Y_1, \ldots, Y_n$ are independently distributed with the common distribution $\bar{\pi}'$, and the marginal distribution of each X_i is $\bar{\pi}$.

(In particular, then, the equality sign in (7.5.2) will hold when $X_1, \ldots, X_n$ are independently distributed with the common distribution $\bar{\pi}$.)

Proof: From (7.5.1) we have

$$H(Y \mid X) = \sum_{i=1}^{n} \sum_{j=1}^{a} g_i(j) H(w(\cdot \mid j)). \tag{7.5.3}$$

Also, from (2.2.14), we have

$$H(Y) \leqq \sum_{i=1}^{n} H(h_i(\cdot)), \tag{7.5.4}$$

with the equality sign holding if and only if $Y_1, \ldots, Y_n$ are independently distributed. It follows from (7.5.3) and (7.5.4) that, for fixed $g_1, \ldots, g_n, h_1, \ldots, h_n$, $H(Y) - H(Y|X)$ would attain its maximum if and only if $Y_1, \ldots, Y_n$ were independent. In particular, the latter will occur when $X_1, \ldots, X_n$ are independent. Temporarily fix $g_1, \ldots, g_n$, $h_1, \ldots, h_n$ at any realizable such system of distributions. Now change (if necessary) the joint distribution of $X_1, \ldots, X_n$ so that their marginal distributions remain unchanged but the X_i are now independently distributed. By the remark just preceding the theorem the marginal distributions of the Y_i are unchanged. On the other hand, the Y_i are now independent, as noted at the end of the last paragraph. It follows therefore from the last paragraph that it remains only to maximize

$$\sum_{i=1}^{n} [H(h_i(\cdot)) - \sum_{j=1}^{a} g_i(j) H(w(\cdot \mid j))]$$

with respect to $g_1, \ldots, g_n$. It is obvious that $g_i = \bar{\pi}$ ($h_i = \bar{\pi}'$) is a solution of this maximization problem. This proves the theorem.

We shall now apply Theorems 7.4.2 and 7.5.1 to the d.m.c. Let $\varepsilon > 0$ be arbitrary, and suppose given any code $(n, 2^{n(C+\varepsilon)}, \lambda)$ for the d.m.c. From the theorems cited we have that

$$\lambda > \frac{n(C+\varepsilon) - nC - 1}{n(C+\varepsilon)} = \frac{\varepsilon}{C+\varepsilon} - \frac{1}{n(C+\varepsilon)} \tag{7.5.5}$$

where C, of course, is the capacity of the d.m.c. Thus we have that, for $\lambda < \frac{\varepsilon}{C+\varepsilon}$ and n sufficiently large, no code $(n, 2^{n(C+\varepsilon)}, \lambda)$ for the d.m.c. exists. However, this is a weaker result than Theorem 3.3.1, which implies that already for *any* $\lambda < 1$, and not only for $\lambda < \frac{\varepsilon}{C+\varepsilon}$, and for n sufficiently large, no code $(n, 2^{n(C+\varepsilon)}, \lambda)$ for the d.m.c. exists.

Again consider the d.m.c., with notation as in Section 3.1. Write $R(\pi)$ for the expression in the left member of (3.1.12). Let $Z(k)$, $k = 1, \ldots, a$, be a chance variable with the distribution

$$P\{Z(k) = j\} = w(j \mid k), \quad j = 1, \ldots, a.$$

Define the chance variable

$$Z^*(k) = \log\left(\frac{w(Z(k) \mid k)}{\bar{\pi}'(Z(k))}\right),$$

where $\bar{\pi}'(Z(k))$ is the $Z(k)^{\text{th}}$ component of the vector $\bar{\pi}'$ (i.e., instead of writing $Z(k)$ as a subscript to $\bar{\pi}'$ we write it as an argument for typographical simplicity). Now

$$R(\pi) = -\sum_{i=1}^{a}\sum_{j=1}^{a} \pi_i w(j \mid i) \log\left(\sum_{i=1}^{a} \pi_i w(j \mid i)\right)$$
$$+ \sum_{i=1}^{a}\sum_{j=1}^{a} \pi_i w(j \mid i) \log w(j \mid i).$$

Differentiate $R(\pi)$ with respect to π_k, $k = 1, \ldots, a$, without regard to the fact that $\sum \pi_i = 1$. We obtain

$$\begin{aligned} \frac{\partial R(\pi)}{\partial \pi_k} &= -\sum_{j=1}^{a} w(j \mid k) \log \pi_j' \\ &+ \sum_{j=1}^{a} w(j \mid k) \log w(j \mid k) - \sum_{j=1}^{a}\sum_{i=1}^{a} \pi_i w(j \mid i) \cdot \left(\frac{w(j \mid k)}{\pi_j'}\right). \end{aligned} \tag{7.5.6}$$

Hence

$$\left.\frac{\partial R(\pi)}{\partial \pi_k}\right|_{\pi=\bar{\pi}} = -1 + EZ^*(k). \tag{7.5.7}$$

Since $R(\pi)$ has a maximum at $\pi = \bar{\pi}$, the left member of (7.5.7) must be a constant, say $K - 1$, for all indices k such that $\bar{\pi}_k > 0$, and must be $\leqq K - 1$ for all indices k such that $\bar{\pi}_k = 0$. Define

$$\begin{aligned} B_1 &= \{k \mid \bar{\pi}_k > 0\} \\ B_2 &= \{k \mid \bar{\pi}_k = 0\}. \end{aligned} \tag{7.5.8}$$

Hence

$$\begin{aligned} EZ^*(k) &= K \quad \text{for } k \varepsilon B_1 \\ EZ^*(k) &\leqq K \quad \text{for } k \varepsilon B_2. \end{aligned} \tag{7.5.9}$$

Now

$$C = \sum_{k=1}^{a} \bar{\pi}_k EZ^*(k). \tag{7.5.10}$$

Hence $K = C$. Thus we have proved

Theorem 7.5.2. For the d.m.c. (7.5.9) holds with $K = C$.

7.6. Discussion of the foregoing. We now discuss the essential difference between the weak and strong converses, say for the d.m.c. Suppose (5.1.1) is a code (n, N, λ). In general, the smaller λ the larger must the A_i be, and the larger λ the smaller may the A_i be. The strong converse puts an upper bound on the number of A_i which can be packed into the space D''_n of the a^n possible received sequences. For arbitrary $\varepsilon > 0$ and sufficiently large n, $2^{n(C+\varepsilon)}$ is such an upper bound, where C is of course the capacity. This bound does not depend on λ, and is therefore the same no matter how big or how small the A_i are. (Of course, the lower bound on n is a function of λ.) The weak converse gives this upper bound only when λ is sufficiently small, i.e., when the A_i are sufficiently large.

For the d.m.c. and for any n and λ let $N^*(n, \lambda)$ be the length of the longest code with probability of error $\leqq \lambda$. Then it follows from Theorems 3.2.1 and 3.3.1 that

$$\lim_{n\to\infty} \frac{1}{n} \log N^*(n, \lambda) = C, \quad 0 < \lambda < 1.$$

That this limit should be the same for all λ (between 0 and 1) is a very interesting phenomenon which enhances the beauty of the whole theory. The use of the word "capacity" without a qualifying λ or other phrase predisposes the reader to regard this fact as obvious and commonplace, but it is not so at all. Once the fact is recognized a plausibility argument can be given for it, but recognition of this fact is not least among the achievements in the founding of the theory. However, this fact can be demonstrated only by proving the strong converse; the weak converse is insufficient to prove it. In the next section we will give a prescription for constructing channels without a capacity. For these channels a weak converse but not a strong converse holds.

It is a common practice in the literature of information theory to refer to a number as the capacity of a channel, although only a weak converse has been proved for the channel. It is to be hoped that this practice will not obscure the need for proving the strong converse where such a need exists.

Let $\lambda(n, N)$ be the minimum probability of error of a code of length N with words of length n. Some writers state Theorems 3.2.1 and 3.3.1, for example, as follows: "For any $R < C$, $\lambda(n, 2^{nR}) \to 0$ (as $n \to \infty$), and for any $R > C$, $\lambda(n, 2^{nR}) \to 1$". It is obvious that each of these statements is weaker than the theorem it purports to state.

A priori it might be expected that the proof of a converse would be more difficult than the proof of a coding theorem. To prove the latter one needs only to exhibit a single code, to prove a converse one has to consider all possible codes. Consequently the weak converse is all the more amazing for its simplicity. It is very striking that its proof uses only

T and not the detailed structure of the channel. By contrast, the proofs of the strong converse, as given in this monograph for various channels, make use of the detailed structure of the channels. The strong converse seems much the deeper result.

The coding theorems of Section 7.2 are results of great generality. It seems to the author that the method of proof used in Theorems 3.2.1 and 4.3.1 lends itself more readily to heuristic purposes. (Of course, this is only an opinion which some may wish to gainsay.) Thus, the results of Section 4.5 and Theorem 4.6.2 are naturally suggested by this method. Its intuitive ideas are easy to understand, and it clarifies the role of the entropy function without any appeal to mysticism or metaphysics. After being led to conjecture or prove a theorem by this method one can try to translate the theorem into the form of Theorem 7.2.1, and then try to adapt the proof of the latter theorem to a proof of the conjecture. The proof of Theorem 3.2.1, for example, shows that a code of the length required can be constructed so that all the u_i's are $\bar{\pi}$-sequences. Perhaps this fact can be of some use in the actual construction of codes.

The number and complexity of possible channels is infinite. However, many channels not considered in our study can be treated by the methods employed here. Sometimes these channels can be reduced by various devices to channels explicitly treated. The methods of various chapters can be combined to give the treatment of new channels.

By combining the methods of Theorems 3.2.1 and 3.3.1 we have obtained a computable value for the capacity of the channel of Section 3.6. (The value in (3.6.3) is not per se a computable one, but Theorem 3.6.3 makes it computable.) It seems very desirable that the value be computable, i.e., that, given any $\varepsilon > 0$, there exist a number $\gamma(\varepsilon)$ such that performing at most $\gamma(\varepsilon)$ prescribed arithmetical operations will yield a number which differs from the capacity by less than ε. The value in (3.1.11) is obviously a computable one. The following is an example of a (per se) non-computable value: "The capacity is the supremum, over all stationary s-inputs, of the rate of transmission of an s-input."

7.7. Channels without a capacity. For any channel let $N(n, \lambda)$ be the length of the longest code of word length n and probability of error λ. Suppose

$$\lim_{n\to\infty} \frac{1}{n} \log N(n, \lambda) = C(\lambda) \tag{7.7.1}$$

exists; let us then call $C(\lambda)$ the λ-capacity of the channel. The definition, for example, of the general discrete channel, as given in Section 5.1, is such that $C(\lambda)$ need not exist. However, $C(\lambda)$ is likely to exist for all λ for most "reasonable" channels; indeed, for the principal channels studied we actually have $C(\lambda) \equiv C$, a constant. We then call C the capacity of the channel. It is obvious that $C(\lambda)$, when it exists, is a monotonically non-decreasing function of λ. When it exists for small λ define

$$C(0+) = \lim_{\lambda \to 0} C(\lambda).$$

Then, for any $\varepsilon > 0$, any λ sufficiently small, and then, depending on λ and ε, any n sufficiently large, we have

$$N(n, \lambda) < 2^{n(C(0+)+\varepsilon)}. \tag{7.7.2}$$

It is this fact which causes some writers to confuse the strong and weak converses.

We now give a particularly simple example of a channel (say D) which does not have a capacity. Suppose given two d.m.c.'s, with respective c.p.f.'s w_1 and w_2, and respective capacities C_1 and C_2, with $C_1 < C_2$. When a word is to be sent over channel D, a chance experiment is performed, with outcomes 1 and 2 and respective non-zero probabilities γ and $1-\gamma$; the experiment is independent of the word to be sent and of all previous experiments. If outcome i ($i = 1, 2$) appears the word to be sent is transmitted over the ith of the d.m.c.'s. The receiver (but not the sender) knows the outcome of the experiment. This completes the description of the channel of which we will now prove that it does not have a capacity.

Suppose $\lambda > \gamma$. Suppose given a code L for the second of the two d.m.c.'s, and let it serve as a code for channel D as follows: The sender uses for transmission of messages the message (transmitted) sequences of the code L. When the word sent is transmitted over the second of the d.m.c.'s, the receiver decodes the word received according to the code L. When the word sent is transmitted over the first of the d.m.c.'s the receiver decodes the word received arbitrarily. We conclude that, for the channel D, when $\lambda > \gamma$ we have

$$\underline{\lim}_{n \to \infty} \frac{1}{n} \log N(n, \lambda) \geq C_2. \tag{7.7.3}$$

Now suppose $\lambda < \gamma$. Suppose, furthermore, that now, when a word is transmitted over the second of the d.m.c.'s, it is correctly understood (decoded) by the receiver with probability one. This assumption certainly does not diminish $N(n, \lambda)$. Yet, obviously, in this case we have

$$\lim_{n \to \infty} \frac{1}{n} \log N(n, \lambda) = C_1. \tag{7.7.4}$$

The relations (7.7.3) and (7.7.4) show that channel D does not have a capacity.

Essentially what is it which causes the capacity to exist? For, as has been remarked earlier, it is the existence of the capacity which is the remarkable fact. Consider the d.m.c. channel and the notation of

Sections 3.1—3.3. Obviously, the smaller λ the more sequences (really the more probability) must there be in each A_i, and the larger λ the smaller may each A_i be. $N(n, \lambda)$ is the maximum number of A_i which can be packed into the space of all n-sequences. Yet the existence of a capacity C implies that, whether λ be large or small, $N(n, \lambda)$ is essentially 2^{nC}, independent of λ.

What, therefore, is it which causes the capacity to exist? Again consider the d.m.c. and the lemmas of Section 2.1. It is easy to see that the capacity exists because the standard deviations of the chance variables which occur in the proofs of the lemmas are of order less than n. Only if these standard deviations were of order n could a capacity fail to exist. In Section III of WOLFOWITZ [7] the reader will find a simple method of modifying the d.m.c. so that one obtains a channel which does not have a capacity. It is immediately apparent from this method that the standard deviations mentioned above become of order n.

7.8. Strong converses. Let D_2 be the totality of all probability distributions on G_2. We now prove

Theorem 7.8.1. Suppose that

$$\min_{d\in D_2} \max_{i\in G_1} \sum_{j\in C_i(d)} h(j|i) < \gamma\,, \tag{7.8.1}$$

where

$$C_i(d) = \{j \in G_2 | h(j|i) \geqq d(j) \cdot 2^{\theta}\}\,. \tag{7.8.2}$$

Suppose that there exists a code (N, λ) with $\lambda < 1 - \gamma$. Then

$$N < (1 - \lambda - \gamma)^{-1} 2^{\theta}. \tag{7.8.3}$$

The proof is very simple. Let d^* be the element of D_2 which minimizes the expression in the left member of (7.8.1). Consider the code (7.1.2), and for any integer k, $1 \leqq k \leqq N$, define $B_k(d^*) \subset G_2$ by

$$B_k(d^*) = A_k \cap \bar{C}_{u_k}(d^*)\,. \tag{7.8.4}$$

Then, from (7.8.1) and (7.8.4) we have that

$$2^{\theta} \cdot \sum_{j\in B_k} d^*(j) > \sum_{j\in B_k} h(j|u_k) > 1 - \lambda - \gamma, \tag{7.8.5}$$

$$k = 1, \ldots, N\,.$$

Summing left and right members with respect to k we obtain that

$$2^{\theta} > N(1 - \lambda - \gamma)\,,$$

which is the desired result.

As an application of this theorem we now prove Theorem 4.8.2. Return to the channel of Section 4.8. Suppose given a code (n, N, λ) for this channel. Suppose that the ith word of the code is to be sent $(i = 1, \ldots, N)$, that the first $(j-1)$ letters of the word have already been sent $(j = 1, \ldots, n)$, and the chance received letters $Z_1, \ldots, Z_{j-1}$ (known to both sender and receiver) have already been received. Then the distribution of the jth chance received letter Z_j, given the above, is the same as that of

$$Z\,(f(i, j, Z_1, \ldots, Z_{j-1}))\,.$$

It follows from Theorem 7.5.2 that

$$E\,(Z_j^* | Z_1, \ldots, Z_{j-1}) \leqq C, \qquad j = 1, \ldots, n \tag{7.8.6}$$

where

$$Z_j^* = \log \left(\frac{w\,(Z_j | f\,(i, j, Z_1, \ldots, Z_{j-1})}{\bar{\pi}'(Z_j)}\right). \tag{7.8.7}$$

Define, for $j = 1, \ldots, n$,

$$W_j = Z_j^* - E\,(Z_j^* | Z_1, \ldots, Z_{j-1})\,.$$

Hence

$$E\,(W_j | Z_1, \ldots, Z_{j-1}) = 0\,. \tag{7.8.8}$$

Therefore

$$E\,(W_j | W_1, \ldots, W_{j-1}) = 0\,, j = 1, \ldots, n \tag{7.8.9}$$

and

$$E\,W_j W_k = 0, j \neq k\,. \tag{7.8.10}$$

Let d^2 be a (finite) upper bound on $E\,W_j^2$ for all values of j; obviously such a bound exists. Let $\xi > 0$ be any number. It follows from (7.8.9), (7.8.10), and CHEBYSHEV's inequality that

$$P\left\{\sum_{j=1}^{n} W_j > \xi \sqrt{n}\right\} \leqq \frac{d^2}{\xi^2}. \tag{7.8.11}$$

From the definition of W_j, (7.8.6) and (7.8.11) we obtain that

$$P\left\{\sum_{j=1}^{n} Z_j^* > nC + \xi \sqrt{n}\right\} \leqq \frac{d^2}{\xi^2}. \tag{7.8.12}$$

Theorem 4.8.2 is now a consequence of (7.8.12) and Theorem 7.8.1.

Of course, Theorem 4.8.2 implies Theorem 3.3.1.

Now consider a channel which is the *product* of two d.m.c.'s. This means the following: A typical word ($2n$-sequence) transmitted is

$u = (u^{(1)}, u^{(2)})$, where the word $u^{(1)}$ (n-sequence) is transmitted over the first d.m.c., and the word $u^{(2)}$ (n-sequence) is transmitted independently over the second d.m.c. The receiver is to decode u from the two sequences transmitted by the first and second d.m.c.'s. Suppose C_1 and C_2 are the respective channel capacities, and let C be the capacity of the product channel (it will be shown to exist).

Obviously $C \geqq C_1 + C_2$, since the errors of decoding $u^{(1)}$ and $u^{(2)}$ separately are at most additive. It remains only to prove that a code for the product channel cannot be longer that $\exp_2\{n(C_1 + C_2) + \sqrt{n}K(\lambda)\}$, where λ is the maximum probability of error and $K(\lambda)$ is a suitable function. If this were not true we would have an information theory perpetuum mobile. To prove the desired result we use the above proof of Theorem 4.8.2 for both d.m.c.'s simultaneously. We employ Theorems 7.5.2 and 7.8.1. Since there is no feedback some of the argument of Theorem 4.8.2 can be omitted. We leave the details to the reader.

Of course, even if there were feedback in the product channel the proof sketched above would go through.

7.9. The strong converse for the d.m.c. revisited. The method to be described in the present section will have wide applicability. It is based on the idea of "blowing up" a decoding set. What this means will be apparent below; the blowing up operation is the repeated application of the operation Γ. The information theoretic application will be discussed after the proof of Theorem 7.9.3.

Unless the contrary is stated, we will adopt the notation of Chapter 3. To eliminate any apprehension about loss of generality, we will henceforth assume that the input alphabet is $\{1,\ldots, a\}$ and the output alphabet is $\{1,\ldots, b\}$. Temporarily fix n. Let v_0 be any n-sequence in the output alphabet. By $\Gamma(v_0)$ we mean the set of all n-sequences which are different from v_0 in at most one element. Let A be a set of n-sequences in the output alphabet. We define

$$\Gamma A = \bigcup_{v_0 \varepsilon A} \Gamma(v_0)\,, \tag{7.9.1}$$

and let $\Gamma^k A$ denote the set of all n-sequences obtained from A by k repetitions of the operation Γ. The boundary ∂A is defined by

$$\partial A = A \cap \Gamma\bar{A}\,, \tag{7.9.2}$$

where $\bar{A}$ is the complement of A. Also define

$$\phi(t) = (2\pi)^{-1/2} \exp_e \left\{-\frac{t^2}{2}\right\}, \tag{7.9.3}$$

$$\Phi(t) = \int_{-\infty}^{t} \phi(s)\,ds, \; f(t) = \phi\,[\Phi^{-1}(t)]\,,$$

where Φ^{-1} is the inverse function of Φ. Let m be the smallest non-zero value of $w(j|i)$, $i = 1,\ldots, a; j = 1,\ldots, b$.

We now note that $f(t)$ is defined for $0 < t < 1, f' = +\sqrt{-2\ln(\sqrt{2\pi}\,f)}$ for $t \leqq \frac{1}{2}$, $f' = -\sqrt{-2\ln(\sqrt{2\pi}\,f)}$ for $t \geqq \frac{1}{2}$, $f\left(\frac{1}{2}\right) = (2\pi)^{-1/2}$, $f'\left(\frac{1}{2}\right) = 0$, $f'' = -\frac{1}{f}$, so that f is symmetric about $\frac{1}{2}$ and concave. Let $\bar{t} = \min(t, 1-t)$. Let us verify that $f(t) \geqq 2(2\pi)^{-1/2}\bar{t}$. This is obviously true at $t = \frac{1}{2}$. The graph of $2(2\pi)^{-1/2}\bar{t}$ as a function of t consists of two chords inscribed in the graph of the concave function f, so that the desired inequality follows. Hence

$$|f'(t)| \leqq 2\sqrt{-\ln \bar{t}}\,. \tag{7.9.4}$$

We note that $f(0) = f(1) = 0$, $f'(0) = \infty$, $f'(1) = -\infty$. Since f is concave, $\frac{f(t)}{t} \uparrow \infty$ as $t \downarrow 0$. Write $z = \Phi^{-1}(t)$. It is well known (e.g., Feller [1], p. 175, Lemma 2) that, for $t > 0$ and small,

$$t \sim \frac{\phi(z)}{|z|}. \tag{7.9.5}$$

Hence

$$f(t) \sim t|z|\,. \tag{7.9.6}$$

As $t \to 0$, $z \to -\infty$, and $\frac{\ln|z|}{|z|} \to 0$. Taking logarithms of both sides of (7.9.5) we conclude that

$$\ln t \sim -\frac{1}{2}z^2, \tag{7.9.7}$$

so that from (7.9.6) we obtain that

$$\frac{f(t)}{t}[-2\ln t]^{-1/2} \uparrow 1 \quad \text{as} \quad t \downarrow 0\,. \tag{7.9.8}$$

We shall also need to prove that

$$f(t) \leqq 8\,t\sqrt{-\ln \bar{t}}\,. \tag{7.9.9}$$

Since f is symmetric about $t = \frac{1}{2}$, it is enough to restrict the argument to $t \leqq \frac{1}{2}$. Differentiating both sides of (7.9.9) and noting that (7.9.9) holds for $t = 0$, we conclude from (7.9.4) that it suffices to prove that

$$2\sqrt{-\ln t} \leqq 8\sqrt{-\ln t} - 4\,(\sqrt{-\ln t})^{-1}.$$

This is always true for $0 \leqq t \leqq \frac{1}{2}$.

We now prove

Theorem 7.9.1. Let A be any set of n-sequences in the output alphabet, and u_0 any n-sequence in the input alphabet. Then

$$P\{v(u_0)\,\varepsilon\,\partial A\} \geqq k n^{-1/2} f\,(P\{v(u_0)\,\varepsilon\,A\}) \tag{7.9.10}$$

where

$$k = \frac{m}{9}(-\ln m)^{-1/2}$$

and m is the smallest non-zero value of w.

Proof: The proof of (7.9.10) will be by induction on n. We first verify it for $n = 1$. Let $g = P\{v(u_0)\,\varepsilon\,A\}$. Then we have to verify that

$$\frac{f(g)}{g} \leqq \frac{1}{k} = \frac{9\sqrt{-\ln m}}{m}. \tag{7.9.11}$$

From (7.9.8) we have

$$\frac{f(g)}{g} \leqq \sqrt{-2\,(\ln g)} \tag{7.9.12}$$

We may always assume that $0 < g < 1$, or the theorem is trivially true. Then $g \geqq m$, and the right member of (7.9.12) is at most $\sqrt{-2\,(\ln m)}$. This is obviously smaller than the right member of (7.9.11).

Assume now that (7.9.10) holds for $1, 2, \ldots, n-1$, and let us prove it for n. Let A_j be the set of $(n-1)$-sequences such that $v_0\,\varepsilon\,A_j \Rightarrow v_0 j\,\varepsilon\,A$. Then A is the disjoint union of sets $A_j j$. Let $u_0(x_1, \ldots, x_n)$ and $u_0' = (x_1, \ldots, x_{n-1})$. Define

$$g_j = P\{v(u_0')\,\varepsilon\,A_j\}$$

$$L = \{j \mid w(j \mid x_n) > 0\}.$$

Now

$$g = \sum_{j \varepsilon L} g_j w(j \mid x_n) \tag{7.9.13}$$

Define

$$\Delta = \left[\min_{j \varepsilon L} g_j, \max_{j \varepsilon L} g_j\right]$$

$$d = \max_{j \varepsilon L} g_j - \min_{j \varepsilon L} g_j$$

Since

$$\partial A \supseteqq \bigcup_j [(\partial A_j) \times j], \tag{7.9.14}$$

we have

$$P\{v(u_0)\,\varepsilon\,\partial A\} \geqq \Sigma w(j|x_n)\, P\{v(u_0')\,\varepsilon\,\partial A_j\}\ . \tag{7.9.15}$$

(The sign $\supset$ in (7.9.14) cannot be omitted. For example, suppose $b = n = 2$, $A = \{(1,1), (2,1)\}$.) Since for any $j_1 \neq j_2$,

$$\partial A \supseteqq [(A_{j_1} - A_{j_2}) \times j_1], \tag{7.9.16}$$

we have

$$P\{v(u_0)\,\varepsilon\,\partial A\} \geqq md\ . \tag{7.9.17}$$

The sign $\supset$ in (7.9.16) cannot be omitted. For example, suppose $b = n = 2$, $A = \{(1,1), (1,2)\}$.)

Suppose first that

$$d \geq \frac{k}{m\sqrt{n}}\, f(g)\,\cdot$$

Then (7.9.10) follows from (7.9.17). Assume therefore that

$$d < \frac{k}{m\sqrt{n}}\, f(g)\,\cdot \tag{7.9.18}$$

From (7.9.15) and the induction hypothesis we have

$$P\{v(u_0)\,\varepsilon\,\partial A\} \geq \frac{k}{\sqrt{n-1}}\, \Sigma w(j|x_n)\, f(g_j). \tag{7.9.19}$$

By Taylor's expansion we have, for $j\,\varepsilon\, L$,

$$f(g_j) = f(g) + (g_j - g)\cdot f'(g) + \frac{1}{2}(g_j - g)^2 \cdot f''(\xi)\,, \tag{7.9.20}$$

where $\xi\,\varepsilon\,\Delta$. Hence

$$\sum_j w(j|x_n)\cdot f(g_j) \geq f(g) - \frac{1}{2}\, d^2 \cdot \max_{\xi\,\varepsilon\,\Delta} |f''(\xi)|\ . \tag{7.9.21}$$

From (7.9.19) and the fact that $f'' = -\frac{1}{f}$, we have

$$P\{v(u_0)\,\varepsilon\,\partial A\} \geq k(n-1)^{-1/2}\left[f(g) - \frac{1}{2n}\frac{k^2}{m^2}\, f^2(g)\cdot\left[\min_{\xi\,\varepsilon\,\Delta} f(\xi)\right]^{-1}\right]\cdot \tag{7.9.22}$$

Let $f(g_0) = \min_{\zeta \varepsilon \Delta} f(\xi), g_0 \varepsilon \Delta$. Then

$$P\{v(u_0) \varepsilon \partial A\} \geq \frac{k}{\sqrt{n}} f(g) \cdot \left[\sqrt{\frac{n}{n-1}} - \frac{k^2 f(g)}{2m^2 f(g_0)\sqrt{n(n-1)}}\right] \cdot \tag{7.9.23}$$

It suffices to prove that the quantity in the bracket of the right member is ≥ 1. This is equivalent to showing that

$$\frac{f(g_0)}{f(g)} \geq \frac{k^2}{m^2} \frac{\sqrt{n} + \sqrt{n-1}}{2\sqrt{n}},$$

and this will follow if we prove that

$$\frac{f(g_0)}{f(g)} \geqq \frac{k^2}{m^2} \cdot \tag{7.9.24}$$

Suppose first that $g \leqq \frac{1}{2} \cdot$ Then $g_0 < g$, and $d > g - g_0$.

From the concavity of f we obtain that

$$\frac{f(g_0)}{f(g)} = \frac{t + g_0}{t + g}, \quad t \geqq 0 \cdot$$

The right member equals

$$1 - \frac{g - g_0}{g + t} > 1 - \frac{d}{g} \cdot$$

Other cases are treated similarly, and we obtain that

$$\frac{f(g_0)}{f(g)} \geqq 1 - \frac{d}{\bar{g}} \cdot \tag{7.9.25}$$

From this and (7.9.18) we obtain

$$\frac{f(g_0)}{f(g)} \geqq 1 - \frac{k f(\bar{g})}{m\sqrt{n}\,\bar{g}} \tag{7.9.26}$$

Since $\bar{g} \geqq m^n$ (recall that we assumed that $0 < g < 1$) we obtain from (7.9.9) and (7.9.26) that

$$\frac{f(g_0)}{f(g)} \geqq 1 - \frac{1}{9\sqrt{-\ln m}\sqrt{n}} \cdot 8\sqrt{-\ln(m^n)} = \frac{1}{9} \cdot$$

On the other hand,

$$\frac{k^2}{m^2} = \frac{1}{81 \ln \frac{1}{m}} < \frac{1}{55}$$

so that (7.9.24) follows and the theorem is proved.

Theorem 7.9.1 enables us to prove the next theorem which is the one with important application to information theory.

Theorem 7.9.2. We have

$$P\{v(u_0)\,\varepsilon\,\Gamma^t A\} \geqq \Phi\left[\Phi^{-1}(P\{v(u_0)\,\varepsilon\,A\} + \frac{1}{\sqrt{n}}(t-1)k\right] \quad (7.9.27)$$

where k is the constant defined in Theorem 7.9.1.

Proof: We note that

$$\Gamma A - A \supseteq \partial(\Gamma A)\,,\ \Gamma A - A = \partial\bar{A}\,. \quad (7.9.28)$$

Write $g_t = \Phi^{-1}(P\{v(u_0)\,\varepsilon\,\Gamma^t A\})$. From the first part of (7.9.28) we obtain

$$\Phi(g_{t+1}) - \Phi(g_t) \geqq P\{v(u_0)\,\varepsilon\,\partial\Gamma^{t+1}A\}$$

and from the second part of (7.9.28) we obtain

$$\Phi(g_{t+1}) - \Phi(g_t) = P\{v(u_0)\,\varepsilon\,\partial(\overline{\Gamma^t A})\}\,.$$

Since f is symmetric about $\frac{1}{2}$ we obtain from Theorem 7.9.1 that

$$\Phi(g_{t+1}) - \Phi(g_t) \geqq \frac{k}{\sqrt{n}}\max\{\phi(g_t),\,\phi(g_{t+1})\}\,.$$

The function ϕ is monotonic from $-\infty$ to 0 and from 0 to ∞. Hence, if $g_t < g_{t+1} \leqq 0$, the maximum of $\phi(u)$ for $g_t \leqq u \leqq g_{t+1}$ is $\phi(g_{t+1})$. Now, by Rolle's theorem applied to the function Φ we have, for some ξ in the interval (g_t, g_{t+1}),

$$\Phi(g_{t+1}) - \Phi(g_t) = (g_{t+1} - g_t)\,\phi(\xi) \leqq (g_{t+1} - g_t)\,\phi(_{t+1})\,,$$

so that $g_{t+1} - g_t \geqq \frac{k}{\sqrt{n}}\cdot$ Since Φ is a monotonically increasing function, (7.9.27) follows. Essentially the same argument applies if $0 \leqq g_t < g_{t+1}$. If 0 lies in the closed interval $[g_t, g_{t+1}]$, we omit the corresponding increment. This completes the proof of the theorem.

We now prove

Theorem 7.9.3. Let $\varepsilon > 0$, $R > 0$, $0 < \lambda' < \lambda < 1$. Let $\{(u_i, A_i)\}$ be a code $(n, \exp(nR), \lambda)$ for a discrete memoryless channel. When n is larger than a bound which depends only on the c.p.f., there exists a subset $\{u_i'\}$ of $\{u_i\}$ whose length is $\exp\{n(R-\varepsilon)\}$, and (in general) new sets A'_i, such that $\{(u_i', A_i')\}$ is a code $(n, \exp[n(R-\varepsilon)], \lambda')$.

Thus every code for the d.m.c. and for sufficiently large n contains a subcode "almost" as long with smaller maximum error. From the given words a subset is selected whose code words are "farther apart", in a certain sense. The new decoding sets are larger and hence the maximum error is smaller. The theorem says that the maximum error can be made arbitrarily small, while the rate is decreased only by ε. However, the reader must note that the new code selects, on the average, one code word out of $\exp(n\varepsilon)$ old code words, so that the ratio of lengths approaches zero.

Before proving the theorem we shall show that it and the weak converse together imply, in the notation of Sections 3.1–3.3, that, for any $\varepsilon' > 0$ and any $\lambda > 0$, no code for the d.m.c. with maximum error λ can have length greater than

$$\exp\{n(C+\varepsilon')\}$$

for n large enough. For suppose there were such codes for infinitely many n. Let $N(n)$ be the length of the one whose words are of length n. Thus

$$N(n) > \exp\{n(C+\varepsilon')\}$$

infinitely often. Now choose the ε of the theorem equal to $\varepsilon'/2$, and the λ' of the theorem such that

$$\frac{C}{1-\lambda'} < C + \frac{\varepsilon'}{4}\cdot$$

Thus we have that there are infinitely many codes

$$\left(n, \exp\left[n\left(C+\frac{\varepsilon'}{2}\right)\right], \lambda'\right)$$

(by the theorem). On the other hand, it follows from the weak converse [(7.4.6)] that the length of a code with maximum error λ' is

$$< \frac{nC+1}{1-\lambda'} < n\left(C+\frac{\varepsilon'}{2}\right)$$

for n sufficiently large. This is a contradiction.

This theorem will not, however, prove the stronger result of Theorem 3.3.1, because as the reader will see in (7.9.32), one must have $k_n(n)^{-1/2} \to \infty$.

Proof of Theorem 7.9.3: We adopt the notation of Sections 3.1–3.3, except that we allow the output alphabet to have b elements. Let $k_n = \sqrt{n}\,(\log n)$. Write $B_i = \Gamma^{k_n}(A_i)$. Since $\frac{k_n}{\sqrt{n}} \to \infty$ we have from Theorem 7.9.2 that, for n sufficiently large,

$$P\{v(u_i)\,\varepsilon\,B_i\} \geqq 1 - \frac{\lambda'}{4},\ i = 1,\ldots,\exp(nR)\,.$$

Let v_n be any n-sequence in the output alphabet. Then the number of sequences in $\Gamma^{k_n}\{v_n\}$ is, for large n,

$$\#\,(\Gamma^{k_n}\{v_n\}) \leqq \binom{n}{k_n}\, b^{k_n} \leqq (nb)^{k_n} \leqq \exp\{2n^{1/2}\log^2 n\}\,. \tag{7.9.30}$$

Hence v_n can belong to at most $\exp\{2n^{1/2}\log^2 n\}$ B_i's.

Let $M = \{2\exp n(R-\varepsilon)\}$. Let $(U_1,\ldots,U_M)$

be independent chance variables, each with the uniform distribution on the integers $1, \ldots, N$ $(N = \exp(nR))$. For typographical simplicity we will write u_i' for u_{U_i}. With each chance variable u_i' associate the (chance) set of n-sequences in the output alphabet defined by

$$D_i = \left\{ v^n \varepsilon B_{U_i} \middle| v^n \varepsilon \bigcup_{j \neq i} B_{U_j} \right\}. \tag{7.9.31}$$

Let $\bar{\lambda}(u_1', \ldots, u_M')$ be the average error of this chance code with decoding sets $D_1, \ldots, D_M$. Then

$$\begin{aligned} E\bar{\lambda}(u_1', \ldots, u_M') &= M^{-1} \sum_{j=1}^{M} E P\{v(u_j') \notin D_{U_j}\} \\ &= E P\{v(u_1') \notin D_{U_1}\} = \\ &= E P\{v(u_1') \notin B_{U_1}\} + E P\{v(u_1') \varepsilon (B_{U_1} - D_{U_1})\}. \end{aligned}$$

The first term of the last member is $< \frac{\lambda'}{4}$, by the construction of the B_i. The probability that a sequence which lies in B_{U_1} will not lie in D_{U_1} is, by (7.9.30), less than

$$1 - \left(1 - \frac{\exp(2n^{1/2}\log^2 n)}{N}\right)^{M-1} \leqq \frac{M}{N} \exp(2n^{1/2}\log^2 n).$$

Hence

$$\begin{aligned} &E P\{v(u_1') \varepsilon (B_{U_1} - D_{U_1})\} \\ &\leqq \frac{M}{N} \exp(2n^{1/2}\log^2 n)\, E P\{v(u_1') \varepsilon B_{U_1}\} \qquad (7.9..32) \\ &\leqq \frac{M}{N} \exp(2n^{1/2}\log^2 n). \end{aligned}$$

Because of the definition of M the right member of (7.9.32) approaches zero as $n \to \infty$. Choose n so large that it is less that $\frac{\lambda'}{4}$. Therefore there then exists a code with average error $< \frac{\lambda'}{2}$ and code length $2\exp\{n(R - \varepsilon)\}$. The theorem now follows from Lemma 3.1.1.

The basic idea of the method of blowing up a decoding set, which can be applied in many cases, can already be seen in the proof of Theorem 7.9.3. By applying k_n times the operation Γ, one can make the size (probability) of each decoding set as close to 1 as desired, while at the same time any sequence in a blown up decoding set belongs (see (7.9.30)) to a number of blown up decoding sets whose ratio to $\exp\{n\varepsilon\}$ approaches zero as $n \to \infty$ (see (7.9.32)).

Remarks

Theorem 7.2.1 is essentially due to FEINSTEIN [1]. Theorem 7.3.2 is due to SHANNON [2]. Theorem 7.2.2 is due to BLACKWELL, BREIMAN, and THOMASIAN [2]. Theorem 7.2.3 is due to the author (WOLFOWITZ [6].) Theorems 7.4.1 and 7.4.2 are due to FANO [1]. Theorem 7.4.3 is due to BLACKWELL, BREIMAN, and THOMASIAN [2]. Threorems 7.4.4 and 7.4.5 are due to the author (WOLFOWITZ [6]). The idea on which the construction of channel D of Section 7.7 is based is due to JACOBS [1]. (See also WOLFOWITZ [7].) Theorem 7.9.1 is credited by AHLSWEDE, GACS, and KÖRNER ([1], [2]) to MARGULIS ([1]). These authors realized the significance of the theorem and proved Theorem 7.9.2, which is the basis of the method of blowing up a decoding set. Theorem 7.9.3 is due to AHLSWEDE and DUECK ([1]).

8. The Semi-Continuous Memoryless Channel

8.1. Introduction. The semi-continuous memoryless channel (s.c.m.c.) differs from the d.m.c. only in that the output alphabet may be infinite. (The name "semi-continuous" is of engineering origin.) Without any essential loss of generality we take the space of the output alphabet to be the real line. This is done only to avoid ponderousness, and the extension to the case where the space of the output alphabet is any space on which is defined a σ-algebra of sets is trivial.

Thus the input alphabet is the set $\{1, \ldots, a\}$, and the c.p.f. consists of a distribution functions (d.f.'s) $F(\cdot \mid 1), \ldots, F(\cdot \mid a)$. If the n-sequence $u_0 = (x_1, \ldots, x_n)$ is sent, the chance received sequence

$$v(u_0) = (Y_1(u_0), \ldots, Y_n(u_0))$$

is a sequence of independent chance variables with distributions given by

$$P\{Y_i(u_0) < z\} = F(z \mid x_i), \quad i = 1, \ldots, n. \tag{8.1.1}$$

Let μ be the measure defined for every Borel set of the real line as the sum of the probabilities assigned to this set by $F(\cdot \mid 1), \ldots, F(\cdot \mid a)$. Let $f(\cdot \mid 1), \ldots, f(\cdot \mid a)$ be, respectively, Radon-Nikodym derivatives of $F(\cdot \mid 1), \ldots, F(\cdot \mid a)$ with respect to μ. Of course these derivatives are determined only to within sets of μ-measure zero. However, we shall adopt some one determination and henceforth speak of "the" derivative. This is done simply to avoid always having to mention the usual set of measure zero, and need cause the reader no anxiety about correctness or rigor. We may, and do, adopt a determination for which

$$f(\cdot \mid 1) + \cdots + f(\cdot \mid a) \equiv 1.$$

By a real n-sequence we shall mean any sequence of n real numbers; the term n-sequence will be retained to designate any sequence of n numbers, each of which is one of $1, \ldots, a$. The symbol p will denote density with respect to an indicated measure; if no measure is indicated the measure μ or the product measure $\mu \cdots \mu$ (whichever applies) is to be understood. We shall write $p\{v(u_0) = v_0\}$ for the density of $v(u_0)$ at v_0. Whatever this notation lacks in strict logicality is more than made up by its suggestiveness and unambiguity. Thus, if $u_0 = (x_1, \ldots, x_n)$ is any n-sequence, and $v_0 = (b_1, \ldots, b_n)$ is any real n-sequence, we have

$$p\{v(u_0) = v_0\} = \prod_{i=1}^{n} f(b_i \mid x_i).$$

Similar symbols with obvious meanings will also be employed. In general, if a capital letter denotes probability, its small letter will denote the corresponding density; thus, q is the density of Q, q'_n is the density of Q'_n etc.

A code (n, N, λ) for the s.c.m.c. (indeed, for any semi-continuous channel) is a set

$$\{(u_1, A_1), \ldots, (u_N, A_N)\},$$

where the u_i are n-sequences, the A_i are disjoint Borel sets of real n-sequences (i.e., each A_i, considered as a set of points in n-space, is a Borel set), and

$$P\{v(u_i) \varepsilon A_i\} \geqq 1 - \lambda, \quad i = 1, \ldots, N.$$

The terminology of Section 5.1 obviously extends to semi-continuous channels. For the s.c.m.c. it will suffice us to employ only sn-inputs in which the components of the chance transmitted n-sequence u are independently distributed with the common distribution π, exactly as in Chapter 3. With such a stochastic input let v be the chance received real n-sequence (with values in D''_n, which is now the space of all real n-sequences). The components of v are then independent, identically distributed chance variables with common density

$$\sum_{j=1}^{a} \pi_j f(\cdot \mid j) = g(\cdot), \quad \text{say.} \tag{8.1.2}$$

The entropy $H(g)$ of g with respect to μ is defined by

$$H(g) = -\int_{-\infty}^{\infty} g(b) \log g(b) \, \mu(db). \tag{8.1.3}$$

This is also to serve as the definition of the entropy of any density g with respect to a positive, bounded, completely additive measure μ on the Euclidean space in which g is defined. (The restriction to a Euclidean space is, as we have already remarked once and will not bother to note again, completely unnecessary.)

We verify easily that the entropy of a density is *not* in general invariant if we replace μ by a similar measure μ' with respect to which μ is absolutely continuous (and of course replace g by the corresponding density g' such that

$$g' = g \cdot \frac{d\mu}{d\mu'}$$

except perhaps on a set of μ' measure zero). For $g(b)\,\mu(db)$ is invariant and g is not in general invariant. This explains the name "entropy with respect to μ". The entropy $H(g^*)$ with respect to μ^* need not be finite for every positive, bounded, completely additive measure μ^* and every density g^* with respect to the latter, as the following simple example shows ($-\infty < b < \infty$):

$$\mu^*(db) = 2^{-b^2} \cdot \frac{1}{1+b^2} \cdot db$$

$$g^*(b) = 2^{b^2}.$$

However, $H(g^*)$ (with respect to μ^*) is always finite if g^* is bounded (above, of course). To prove this we first observe that it is sufficient to take the bound to be 1, since μ^* assigns a finite measure to the entire space. Now let $B_i = \{b \mid 2^{-i-1} < g^*(b) \leqq 2^{-i}\}$, $i = 0, 1, \ldots$ Then

$$\int_{B_i} g^*(b)\,\mu^*(db) \leqq 2^{-i}\,\mu^*(B_i) \leqq c \cdot 2^{-i}$$

where $c = \sum_i \mu^*(B_i) < \infty$. It follows that $H(g^*)$ with respect to μ^* is dominated by

$$c t_1 = c \cdot \sum_i \frac{i+1}{2^i} < \infty, \tag{8.1.4}$$

as was to be proved. Incidentally, (8.1.3) is finite for g given by (8.1.2), because of what was just proved. Moreover, we have shown the following: Let x^- denote $\min(0, x)$. Then

$$-\int_{-\infty}^{\infty} g(b) \log^- g(b)\,\mu(db) = \overline{H}(g) \text{ (say) } \leqq t_1,$$

where t_1 is the (positive) absolute constant defined in (8.1.4).

Write, for $k > 0$,

$$t_k = \sum_i \frac{(i+1)^k}{2^i}$$

A trivial modification of the argument of the last paragraph gives the following result: Let μ^* and g^* be as in that paragraph, and suppose $g^* \leqq 1$. Then, for $k > 0$,

$$\int g^*(b)\,|\log g^*(b)|^k \mu^*(db) \leqq c t_k. \tag{8.1.5}$$

If it were not for the dependence on μ, $H(g)$ would be the complete analogue of $H(\pi')$ and of $H(Q_1'')$. The dependence on μ is inescapable. Let us therefore reconcile ourselves to the inevitable, adopt the particular μ above, and give the analogues of the other entropies. It seems appropriate to call $H(g)$ the entropy of the output (or of the $s1$-output) w.r. to μ. We may also write it as $H(Q_1'')$. The entropy of the input (of the $s1$-input) is, exactly as before, $H(\pi)$, and obviously does not depend on μ. We may also write it as $H(Q_1')$. The analogues of the other entropies w.r. to μ and with names corresponding to those of Section 5.1, are:

analogue of $H(Q_1'' \mid Q_1')$, the conditional entropy of the $s1$-output, given the $s1$-input, is (8.1.6)

$$-\sum_{i=1}^{a} \pi_i \int_{-\infty}^{\infty} f(b \mid i) \log f(b \mid i)\, \mu(db)$$

analogue of $H(Q_1' \mid Q_1'')$, the conditional entropy of the $s1$-input, given the $s1$-output, is (8.1.7)

$$-\sum_{i=1}^{a} \pi_i \int_{-\infty}^{\infty} f(b \mid i) \log \left(\frac{\pi_i f(b \mid i)}{g(b)}\right) \mu(db).$$

These entropies can also be defined with respect to different measures, provided they (the entropies) are finite. They are certainly finite for our particular μ, by the argument of the preceding paragraph. (To check this for $H(Q_1' \mid Q_1'')$ we note that the latter can be written as

$$\begin{aligned} &-\sum_{i=1}^{a} \int_{-\infty}^{\infty} \pi_i f(b \mid i) \log (\pi_i f(b \mid i))\, \mu(db) \\ &+ \int_{-\infty}^{\infty} g(b) \log g(b)\, \mu(db).) \end{aligned} \tag{8.1.8}$$

One verifies easily that

$$H(Q_1') - H(Q_1' \mid Q_1'') = H(Q_1'') - H(Q_1'' \mid Q_1'). \tag{8.1.9}$$

This is simply the analogue of (5.1.3) for $n = 1$. We say "analogue" because three of the four quantities which appear in (8.1.9) seem to depend on μ. Actually we need not have said this, because the left and right members of (8.1.9) do not depend on μ, and are the same for every μ for which all the terms are defined. To see this it is sufficient to prove that $H(Q_1' \mid Q_1'')$ also does not depend on μ. It is actually the only other entropy, besides $H(Q_1')$, which does not depend on μ. To prove the first statement, that $H(Q_1' \mid Q_1'')$ does not depend on μ, we observe that $f(b \mid i)\, \mu(db)$ is invariant under change of measure, and so is $\frac{\pi_i f(b \mid i)}{g(b)}$.

From (8.1.9) and the non-invariance of $H(Q_1'')$ we conclude that $H(Q_1'' \mid Q_1')$ is not invariant. Another proof of the latter consists in noting that $f(b \mid i)\,\mu(db)$ is invariant under change of μ, but $f(b \mid i)$ and hence $\log f(b \mid i)$ is not in general invariant.

8.2. A coding theorem and its strong converse. As in the case of the d.m.c., we define the quantity C of the s.c.m.c. as the maximum (easily shown to exist) with respect to π, of either member of (8.1.9) (for any μ for which the various entropies are finite). It is easy to see that $C = 0$ if, and only if, $F(\cdot \mid 1) \equiv \cdots \equiv F(\cdot \mid a)$. Let $\bar{\pi}$ be a value of π for which C is achieved.

Let π be any probability a-vector, and ν be the measure defined for any Borel set S of the real line by $\nu(S) = \sum \pi_i F(S \mid i)$. Let $m(\cdot \mid i)$ be a Radon-Nikodym derivative of $F(\cdot \mid i)$ with respect to ν, such that $\Sigma \pi_i m(\cdot \mid i)$ is identically one. Then

$$\begin{aligned}
&\sum_i \pi_i \int_{-\infty}^{\infty} m(b \mid i) \left[\log \frac{m(b \mid i)}{\sum_j \pi_j m(b \mid j)}\right]^2 \nu(db) \\
&= \int \sum_i \pi_i m(b \mid i) \left[\log \frac{\pi_i\, m(b \mid i)}{\pi_i}\right]^2 \nu(db) \\
&\leqq \int \sum_i \pi_i m(b \mid i)\, [\log(\pi_i m(b \mid i))]^2 \nu(db) \\
&+ \int \sum_i \pi_i m(b \mid i)\, [\log \pi_i]^2 \nu(db) \qquad (8.2.1) \\
&= \int \sum_i \pi_i m(b \mid i)\, [\log(\pi_i m(b \mid i))]^2 \nu(db) \\
&+ \sum_i \pi_i\, [\log \pi_i]^2 .
\end{aligned}$$

Since $\sum_i \pi_i m(b \mid i) \equiv 1$, by the argument which follows (7.2.13), the left member of (8.2.1) is $\leqq 2(\log a)^2$ when $a \geqq 3$, and $\leqq 4.686$ when $a = 2$.

It is now a simple matter to prove

Theorem 8.2.1. Let $0 < \lambda < 1$ be arbitrary. There exists a positive function $K(\lambda)$ of λ such that, for any s.c.m.c. and any n, there exists a code

$$(n, \exp_2\{nC - \sqrt{n}K(\lambda) \log a\}, \lambda) \qquad (8.2.2)$$

and there does not exist a code

$$(n, \exp_2\{nC + \sqrt{n}K(\lambda) \log a\}, \lambda)\,. \qquad (8.2.3)$$

(Of course, the function K does not depend on a.)

In proving this result we make use of the conclusion after (8.2.1). We prove (8.2.2) by using the s.c.m.c. analogue of (7.2.13). We prove (8.2.3) by using Theorem 7.8.1 and the s.c.m.c. analogue of Theorem 7.5.1 (which is obviously valid).

For semicontinuous channels other than the s.c.m.c. see WOLFOWITZ [4] and KESTEN [1].

9. Continuous Channels with Additive Gaussian Noise

9.1. A continuous memoryless channel with additive Gaussian noise For brevity we shall henceforth refer to the channel to be studied in the present section as the channel G. Let σ^2 be the variance of the Gaussian noise. We will take the input alphabet of channel G to be the interval $[0, 1]$. In place of (8.1.1) we now have

$$P\{Y_i(u_0) < z + x_i\} = \frac{1}{\sqrt{2\pi}\,\sigma} \int_{-\infty}^{z} \exp_e\left\{\frac{-t^2}{2\sigma^2}\right\} dt \qquad (9.1.1)$$
$$i = 1, \ldots, n.$$

We have $0 \leqq x_i \leqq 1$, $i = 1, \ldots, n$, and the chance variables $Y_1(u_0), \ldots, Y_n(u_0)$ are independently distributed. The definition of a code is obvious.

Let $u_0 = (x_1, \ldots, x_n)$ and $u_0' = (x_1', \ldots, x_n')$ be two real n-sequences such that

$$\frac{-1}{2\gamma\sqrt{n}} \leqq \Delta_i = x_i' - x_i \leqq \frac{1}{2\gamma\sqrt{n}}, \quad i = 1, \ldots, n,$$

where $\gamma > 0$. Let $y = (y_1, \ldots, y_n)$ be any real n-sequence. Then, in the notation of Chapter 8,

$$p\{v(u_0) = y\} = \left(\frac{1}{\sqrt{2\pi}\,\sigma}\right)^n \exp_e\left\{\frac{-\Sigma(y_i - x_i)^2}{2\sigma^2}\right\}, \qquad (9.1.2)$$

$$p\{v(u_0') = y\} = \left(\frac{1}{\sqrt{2\pi}\,\sigma}\right)^n \exp_e\left\{\frac{-\Sigma(y_i - x_i')^2}{2\sigma^2}\right\}. \qquad (9.1.3)$$

Hence

$$\frac{p\{v(u_0') = y\}}{p\{v(u_0) = y\}} > \exp_e\left\{\frac{-|\Sigma(y_i - x_i)\Delta_i|}{\sigma^2} - \frac{1}{8\gamma^2\sigma^2}\right\}. \qquad (9.1.4)$$

Let A be a set of real n-sequences such that

$$P\{v(u_0)\ \varepsilon\ A\} \geqq 1 - \lambda \quad (\lambda < 1). \qquad (9.1.5)$$

From A delete all sequences y such that

$$\left|\sum_{i=1}^{n} (y_i - x_i)\,\Delta_i\right| \geqq \frac{\sigma}{\gamma\sqrt{1-\lambda}}; \qquad (9.1.6)$$

call the resulting set A'. Then, by CHEBYSHEV's inequality,

$$P\{v(u_0)\ \varepsilon\ (A - A')\} \leqq \frac{(1-\lambda)}{4}. \qquad (9.1.7)$$

Now choose γ so large, say $= \gamma_0(\lambda)$, that

$$\exp_e\left\{-\frac{1}{\gamma\sigma\sqrt{1-\lambda}} - \frac{1}{8\gamma^2\sigma^2}\right\} > \frac{2}{3}. \tag{9.1.8}$$

We conclude from (9.1.4) and (9.1.7) that

$$P\{v(u_0') \varepsilon A'\} \geqq \frac{1-\lambda}{2}. \tag{9.1.9}$$

Let a be a positive integer and

$$B(a) = \left\{\frac{i-\frac{1}{2}}{a},\ i = 1, \ldots, a\right\}.$$

Let $C(a)$ be the capacity of the s.c.m.c. with input alphabet $B(a)$ and additive Gaussian noise with variance σ^2. Let

$$\{(u_1, A_1), \ldots, (u_N, A_N)\}$$

be a code (n, N, λ) for channel G $(0 < \lambda < 1)$. We replace $u_i, i = 1, \ldots, n$, by u_i', an n-sequence of elements of $B(\gamma\sqrt{n})$ such that corresponding elements of u_i and u_i' differ by not more than $\frac{1}{2\gamma\sqrt{n}}$ in absolute value. (We assume that $\gamma\sqrt{n}$ is an integer. If this is not so, trivial changes below will suffice to take care of this.) Hence from the previous paragraph it follows that

$$\{(u_1', A_1), \ldots, (u_N', A_N)\}$$

is a code $\left(n, N, \frac{1+\lambda}{2}\right)$ for channel G. It follows from (8.4.14) that, for $\gamma\sqrt{n} \geqq 3$,

$$N < (n+1)^{\gamma\sqrt{n}} \cdot \frac{4}{1-\lambda} \exp_2\left\{nC\left(\gamma\sqrt{n}\right) + 2\sqrt{\frac{n}{1-\lambda}} \log\left(\gamma\sqrt{n}\right)\right\}. \tag{9.1.10}$$

On the other hand, it follows from the paragraph which contains (7.2.13) and the paragraph which follows (8.4.5) that, for any n, there exists a code (n, N, λ) for the s.c.m.c. with input alphabet $B(a)$ such that

$$N > \frac{\lambda}{2} \exp_2\left\{n\,C(a) - t\sqrt{\frac{2n}{\lambda}} \log a\right\}, \tag{9.1.11}$$

where t is the absolute positive constant defined in Section 8.4.

Write

$$C = \sup_a C(a).$$

If $C \neq \infty$ it follows from (9.1.10) and (9.1.11) that

$$\lim_{a\to\infty} C(a) = C. \tag{9.1.12}$$

If $C=\infty$ then (9.1.12) is clearly true. We shall now show that $C<\infty$ and bound $C-C(a)$ when

$$a \geqq \max\left(4, \gamma_0\left(\frac{1}{2}\right)\right) = \gamma_1 \text{ (say)}. \tag{9.1.13}$$

Let a be any integer which satisfies (9.1.13) and let $b>a$ be another integer. Let n be the largest integer such that $\gamma = \frac{a}{\sqrt{n}} \geqq \gamma_1$. Then

$$\gamma_1 \leqq \gamma \leqq \gamma_1 \sqrt{2}. \tag{9.1.14}$$

From (9.1.10) and (9.1.11) for $\lambda = \frac{1}{2}$ we obtain

$$\begin{aligned} &\frac{1}{4} \exp_2 \left\{n\, C(b) - 2t\sqrt{n} \log b\right\} \\ &< 8(n+1)^a \exp_2 \left\{n\, C(a) + 2\sqrt{2n}\, \log a\right\}. \end{aligned} \tag{9.1.15}$$

Hence, putting $n = \frac{a^2}{\gamma^2}$, we obtain

$$\begin{aligned} &a\,(C(b) - C(a)) < \gamma^2 \left(\frac{5}{a} - 2\log\gamma\right) \\ &+ \left(3\gamma^2 + \gamma\sqrt{8}\right) \log a + 2\,\gamma\, t \log b. \end{aligned} \tag{9.1.16}$$

Since $a \geqq \gamma \geqq \gamma_1 \geqq 4$ the first term of the right member is negative. Now replace a by a^{k-1} and b by a^k. Then, from (9.1.14),

$$C(a^k) - C(a^{k-1}) \leqq \frac{d \cdot \gamma_1^2 \cdot k \log a}{a^{k-1}} \tag{9.1.17}$$

where we may take

$$d = (7+t). \tag{9.1.18}$$

Consequently, employing (9.1.17) for $k = 2, 3, \ldots$, we obtain

$$C - C(a) \leqq d\gamma_1^2 \left(\frac{\log a}{a}\right) \left(2 + \frac{3}{a} + \frac{4}{a^2} + \cdots\right). \tag{9.1.19}$$

The inequality (9.1.19) enables us to compute C to any specified accuracy.

We have also proved

Theorem 9.1.1. Let $0<\lambda<1$ and $\varepsilon>0$ be arbitrary. For channel G and all n sufficiently large there exists a code $(n, 2^{n(C-\varepsilon)}, \lambda)$, and there does not exist a code $(n, 2^{n(C+\varepsilon)}, \lambda)$.

9.2. Message sequences within a suitable sphere. The channel we shall now study is such that, for any n, an n-sequence (word) to be transmitted is a point $u_0 = (x_1, \ldots, x_n)$ in the interior or on the periphery of the n-sphere with center at the origin O and radius $\sqrt{n}$. Thus

$$\sum_1^n x_i^2 \leqq n. \tag{9.2.1}$$

When the sequence u_0 is sent over the channel, the received sequence

$$v(u_0) = (Y_1(u_0), \ldots, Y_n(u_0))$$

is a sequence of independent chance variables, and the distribution of $Y_i(u_0)$, $i = 1, \ldots, n$, is normal, with mean x_i and variance σ^2. We shall show that the capacity C_1 of this channel is given by

$$C_1 = \frac{1}{2} \log \frac{1+\sigma^2}{\sigma^2}. \tag{9.2.2}$$

We shall do this, of course, by proving a coding theorem and its strong converse.

To prove a coding theorem we notice that Theorem 7.2.1 holds even when q' is a density on the set G_1 and $w(\cdot \mid i)$ is a density on the set G_2 (for every i). (Obvious measurability conditions must be fulfilled.) Moreover, the corollary to Theorem 7.2.1 applies in this case with the obvious changes. We shall apply this corollary with these changes to the present channel to obtain a coding theorem.

Suppose

$$u = (X_1, \ldots, X_n),$$

and the X_i are independent, identically and normally distributed chance variables, with mean zero and variance $1-\varepsilon'$, where $\varepsilon' > 0$ is a small arbitrary number. Since

$$\frac{X_1^2 + \cdots + X_n^2}{n} \tag{9.2.3}$$

converges stochastically to $1-\varepsilon'$, it follows that, for sufficiently large n, the probability that u should fall into the sphere (9.2.1) is $> 1-\varepsilon'$.

Let q' be the density of u, q'' be the density of $v(u)$, and let q be the density of $(u, v(u))$. Let $v_0 = (y_1, \ldots, y_n)$. At the point (u_0, v_0) we have

$$q'(u_0) = \left(\sqrt{2\pi(1-\varepsilon')}\right)^{-n} \exp\left\{-\frac{\Sigma x_i^2}{2(1-\varepsilon')}\right\}, \tag{9.2.4}$$

$$q''(v_0) = \left(\sqrt{2\pi(1-\varepsilon'+\sigma^2)}\right)^{-n} \exp\left\{-\frac{\Sigma y_i^2}{2(1-\varepsilon'+\sigma^2)}\right\}, \tag{9.2.5}$$

$$q(u_0, v_0) = \left(2\pi\sigma\sqrt{1-\varepsilon'}\right)^{-n} \exp\left\{-\frac{\Sigma x_i^2}{2(1-\varepsilon')} - \frac{\Sigma(y_i - x_i)^2}{2\sigma^2}\right\}. \tag{9.2.6}$$

Hence, as $n \to \infty$,

$$\frac{1}{n} \log \frac{q(u, v(u))}{q'(u)\, q''(v(u))} \tag{9.2.7}$$

converges stochastically to

$$\frac{1}{2} \log \left(\frac{1-\varepsilon'+\sigma^2}{\sigma^2}\right). \tag{9.2.8}$$

We now employ the corollary to Theorem 7.2.1 and obtain the following theorem when ε' is taken sufficiently small:

Theorem 9.2.1. Let λ and ε be arbitrary positive numbers, with $\lambda \leqq 1$. When n is sufficiently large there exists a code (n, N, λ) for the channel of the present section with

$$N > 2^{n(C_1 - \varepsilon)}. \tag{9.2.9}$$

We now prove the strong converse. Let $u_{00} = (c, 0, 0, \ldots, 0)$, $0 \leqq c \leqq \sqrt{n}$, and suppose (u_{00}, A_{00}) is an element of a code (n, N, λ), where n is sufficiently large (how large will appear shortly) and $\lambda < 1$. Let d be a positive number, and delete from A_{00} all points whose first coordinate is either $> c + d$ or $< c - d$; call the resulting set A'_{00}. If d is sufficiently large (which we will henceforth take to be the case) then

$$P\{v(u_{00})\, \varepsilon\, A'_{00}\} > \frac{1-\lambda}{2}. \tag{9.2.10}$$

Fix $\varepsilon' > 0$. From A'_{00} delete all points whose distance from the line from u_{00} to O is greater than $\sigma(1 + \varepsilon') \sqrt{n}$; call the resulting set A''_{00}. Since

$$\frac{(Y_2(u_{00}))^2 + \cdots + (Y_n(u_{00}))^2}{n} \tag{9.2.11}$$

converges stochastically to σ^2 as $n \to \infty$, it follows that, for n sufficiently large,

$$P\{v(u_{00})\, \varepsilon\, A''_{00}\} > \frac{1-\lambda}{4}. \tag{9.2.12}$$

Every point in A''_{00} is at a distance at most

$$\left[\left(\sqrt{n} + d\right)^2 + \left(\sigma\,(1 + \varepsilon') \sqrt{n}\right)^2\right]^{\frac{1}{2}} = \sqrt{n[1 + \sigma^2(1 + \varepsilon'')^2]}\,, \text{ say,} \tag{9.2.13}$$

from O. Obviously $\varepsilon'' \to \varepsilon'$ as $n \to \infty$.

Let

$$\{(u_1, A_1), \ldots, (u_N, A_N)\} \tag{9.2.14}$$

be a code (n, N, λ), with $\lambda < 1$ and n sufficiently large. For $i = 1, \ldots, N$ perform the following operation: Rotate the coordinate axes so that u_i lies on the linefrom O to $\left(\sqrt{n}, 0, 0, \ldots, 0\right)$. With u_i in the role of u_{00}, obtain A''_i from A_i in the same manner as A''_{00} was obtained from A_{00}. Then rotate the coordinate axes back to their original position. The system

$$\{(u_1, A''_1), \ldots, (u_N, A''_N)\} \tag{9.2.15}$$

is a code $\left(n, N, \frac{3+\lambda}{4}\right)$.

The volume of the unit n-sphere is

$$L_n = \frac{\pi^{\frac{n}{2}}}{\Gamma\left(\frac{n+2}{2}\right)}. \tag{9.2.16}$$

Let $\varepsilon_0 > 0$ be arbitrary. From the stochastic convergence of (9.2.11) it follows that, when n is large, the volume of A''_{00} is at least that of a sphere in n-space with radius $\sigma(1-\varepsilon_0)\sqrt{n}$. Hence, when n is large, the volume of $A''_1 \cup \cdots \cup A''_N$ is at least

$$N \cdot L_n \cdot \left(\sigma(1-\varepsilon_0)\sqrt{n}\right)^n. \tag{9.2.17}$$

On the other hand, the volume of $A''_1 \cup \cdots \cup A''_N$ is, when n is large, not greater than the volume of all points whose distance from the origin is (9.2.13). This volume is

$$L_n \cdot \left(\sqrt{n[1+\sigma^2(1+\varepsilon'')^2]}\right)^n. \tag{9.2.18}$$

Now let ε_0 and ε' be suitably small and n sufficiently large. From (9.2.17) and (9.2.18) we obtain

Theorem 9.2.2. Let λ and ε be arbitrary numbers, with $0 \leqq \lambda < 1$ and $\varepsilon > 0$. Let (9.2.14) be a code (n, N, λ) for the channel of the present section. If n is sufficiently large then

$$N < 2^{n(C_1+\varepsilon)}. \tag{9.2.19}$$

9.3. Message sequences on the periphery of the sphere or within a shell adjacent to the boundary. Let $\varepsilon_1 > 0$ be arbitrary. Consider the channel of the preceding section with only the following modification: If u_0 is any word to be transmitted, then, instead of satisfying (9.2.1), it satisfies

$$n(1-\varepsilon_1) \leqq \Sigma x_i^2 \leqq n. \tag{9.3.1}$$

It follows from the corollary to Theorem 7.2.1 and the proof of Theorem 9.2.1 that the latter theorem holds verbatim for our present channel. Since a code for the present channel with probability of error $\leqq \lambda$ cannot be longer than the longest code with probability of error $\leqq \lambda$ for the channel of the preceding section, Theorem 9.2.2 holds verbatim for the present channel a fortiori.

Now consider the channel of Section 9.2 with only the following modification: Relation (9.2.1) is replaced by

$$\sum_1^n x_i^2 = n. \tag{9.3.2}$$

Of course, Theorem 9.2.2 again holds verbatim for the present channel. In order to prove a coding theorem, consider the channel where the words have $(n-1)$ letters, subject to the condition

$$\sum_1^{n-1} x_i^2 \leqq n-1, \tag{9.3.3}$$

and the distribution of the received letters is as throughout this chapter. It follows from Theorem 9.2.1 that for this last channel there exists a code with probability of error $\leqq \lambda$ and length $2^{n(C_1-\varepsilon)}$ ($\lambda > 0$ and $\varepsilon > 0$ arbitrary and n sufficiently large). Now add a letter x_n so that (9.3.2) holds, and ignore the letter $Y(x_n)$. (What this means is obvious.) We have that Theorem 9.2.1 holds verbatim for the channel which is subject to the restriction (9.3.2).

9.4. Another proof of Theorems 9.2.1 and 9.2.2. The results of Sections 9.2 and 9.3 are due to SHANNON [4], who proved them by different methods, adaptations of which we shall now describe. The proof of Theorem 9.2.1 will be by the method of random codes of Section 7.3.

First we prove the strong converse for the channel with restriction (9.3.2). Suppose $u_{000} = (\sqrt{n}, 0, 0, \ldots, 0)$ is one of the elements $u_1, \ldots, u_N$ of (9.2.14). Let θ_1 be the half-angle of a right "circular" cone B_1 with apex at 0, axis the infinite half-line from O through u_{000}, and such that

$$P\{v(u_{000})\,\varepsilon\, B_1\} = 1 - \lambda. \tag{9.4.1}$$

Let $b(\theta_1)$ be the area cut out by this cone from the surface of the unit n-sphere. The surface area of this sphere is

$$M_n = \frac{n\pi^{\frac{n}{2}}}{\Gamma\left(\frac{n+2}{2}\right)}. \tag{9.4.2}$$

Obviously therefore

$$N < \frac{M_n}{b(\theta_1)}. \tag{9.4.3}$$

Let $Y_1, \ldots, Y_n$ be independent, normally distributed chance variables with means zero and variances σ^2. Then, from (9.4.1),

$$\begin{aligned} 1 - \lambda &= P\left\{\frac{Y_1 + \sqrt{n}}{\sqrt{\sum_2^n Y_i^2}} \geqq \cot\theta_1\right\} \\ &= P\left\{\frac{Y_1 + \sqrt{n}}{\sqrt{\frac{\sum_2^n Y_i^2}{n}}} \geqq \sqrt{n}\cot\theta_1\right\} \qquad (9.4.4) \\ &= F(\theta_1), \quad \text{say}. \end{aligned}$$

In general, define

$$F(\theta) = P\left\{\frac{Y_1 + \sqrt{n}}{\sqrt{\sum_2^n Y_i^2}} \geqq \cot\theta\right\}.$$

Let $\varepsilon_1 > 0$ be arbitrary. If, for n large,

$$\sin\theta_1 \leqq \frac{\sigma}{\sqrt{\sigma^2+1}} - \varepsilon_1, \tag{9.4.5}$$

then, since

$$\frac{\sum_{2}^{n} Y_i^2}{n} \tag{9.4.6}$$

converges stochastically to σ^2, all the members of (9.4.4) except the first would approach zero as $n \to \infty$. Hence (9.4.5) cannot hold for large n. Consequently, for n sufficiently large,

$$\sin \theta_1 > \frac{\sigma}{\sqrt{\sigma^2 + 1}} - \varepsilon_1 . \tag{9.4.7}$$

The area $b(\theta)$ cut out from the surface of the unit n-sphere by a cone of half-angle θ is, by an elementary formula, given by

$$b(\theta) = \frac{(n-1)\,\pi^{\frac{n-1}{2}}}{\Gamma\left(\frac{n+1}{2}\right)} \int_0^\theta (\sin x)^{n-2}\, dx . \tag{9.4.8}$$

For any Δ such that $0 < \Delta < \theta$ we have

$$\frac{\Delta (n-1)\,\pi^{\frac{n-1}{2}} [\sin(\theta - \Delta)]^{n-2}}{\Gamma\left(\frac{n+1}{2}\right)} < b(\theta) < \frac{(n-1)\,\pi^{\frac{n+1}{2}} [\sin \theta]^{n-2}}{\Gamma\left(\frac{n+1}{2}\right)} \tag{9.4.9}$$

if $\theta < \frac{\pi}{2}$. The desired result (9.2.19) now follows easily from (9.4.2), (9.4.3), (9.4.7), and (9.4.9), for a Δ suitably small and all n sufficiently large. This proves the strong converse for the channel with restriction (9.3.2).

In order to prove Theorem 9.2.2 (i.e., replace restriction (9.3.2) by (9.2.1)) we simply apply the argument described in Section 9.3.

We now prove Theorem 9.2.1. We may replace restriction (9.2.1) by (9.3.2) in the course of the proof; Theorem 9.2.1 will then be true a fortiori. Suppose $u_1^*, \ldots, u_N^*$ are chance variables distributed independently and uniformly on the surface (9.3.2). Let A_i^*, $i = 1, \ldots, N$, be the set of all points v_0 in n-space such that the angle $v_0 O u_i^*$ is less than angles $v_0 O u_1^*, \ldots, v_0 O u_{i-1}^*, v_0 O u_{i+1}^*, \ldots, v_0 O u_N^*$. Fix $\varepsilon > 0$. Define $N = 2^{n\left(C_1 - \frac{\varepsilon}{2}\right)}$. We will show that

$$E\left[\frac{1}{N} \sum_{i=1}^{N} P\{v(u_i^*) \notin A_i^*\}\right] \to 0 \tag{9.4.10}$$

as $n \to \infty$. Suppose this has been proved. Let n be so large that the left member of (9.4.10) is less than $\frac{\lambda}{2}$. It follows that there exists a value $(u_1^0, \ldots, u_N^0)$ of $(u_1^*, \ldots, u_N^*)$, with the corresponding sets $(A_1^0, \ldots, A_N^0)$

of $(A_1^*, \ldots, A_N^*)$, such that

$$\frac{1}{N}\sum_{i=1}^{N} P\{v(u_i^0) \notin A_i^0\} < \frac{\lambda}{2}. \tag{9.4.11}$$

The desired result (9.2.9) now follows from Lemma 3.1.1.

It remains to prove (9.4.10). Because of the symmetry of the sphere and the distribution of $Y_1, \ldots, Y_n$, it will suffice to prove that

$$E\,P\{v(u_1^*) \in A_1^* \mid u_1^* = u_{000}\} \to 1. \tag{9.4.12}$$

Let γ be the size of the chance angle $u_{000}\,O\,v(u_{000})$. Then, when $\gamma = g$, the event in curly brackets in (9.4.12) will occur if the angles $u_2^*\,O\,v(u_{000})$, $\ldots$, $u_N^*\,O\,v(u_{000})$ are all greater than g. Since $u_1^*, u_2^*, \ldots, u_N^*$ are independently and uniformly distributed on the surface (9.3.2), we have that the left member of (9.4.12) is equal to

$$\int_0^{\pi}\left(1-\frac{b(g)}{b(\pi)}\right)^{N-1} dF(g). \tag{9.4.13}$$

Let $\Delta > 0$ be chosen arbitrarily. We have already seen (argument following (9.4.6)) that, for large n, F assigns probability greater than $1-\Delta$ to an interval I of width 2Δ centered at θ_0, where

$$\sin\theta_0 = \frac{\sigma}{\sqrt{1+\sigma^2}}. \tag{9.4.14}$$

To prove (9.4.12) it will suffice to show that

$$\min_{g\in I}\left(1-\frac{b(g)}{b(\pi)}\right)^{N} \to 1 \tag{9.4.15}$$

as $n\to\infty$. To prove (9.4.15) it will suffice to prove that, as $n\to\infty$,

$$\frac{N\,b(\theta_0+\Delta)}{b(\pi)} \to 0. \tag{9.4.16}$$

Let $\varepsilon' > 0$ and

$$N = \left(\frac{1+\sigma^2}{\sigma^2} - \varepsilon'\right)^{\frac{n}{2}}. \tag{9.4.17}$$

Then (9.4.16) follows at once from (9.4.2) and (9.4.9), if we take Δ to be a suitably small function of ε'. This proves Theorem 9.2.1.

Remarks

Theorem 9.1.1 is due to J. H. B. KEMPERMAN. Theorems 9.2.1 and 9.2.2, as well as the results of Section 9.3, are due to SHANNON [*4*]. The proofs of Section 9.2 are new.

10. Mathematical Miscellanea

10.1. Introduction. This chapter is simply a concession to the conventional treatment of the subject. It contains the proofs of two results (Theorems 10.2.1 and 10.3.1) which are basic in published treatments of the subject, but of which no use was made in this monograph. The discussion of Section 5.6 describes the use to which these results are put. A comparison with the methods of Chapter 5 may be instructive.

10.2. The asymptotic equipartition property. Let D' be the space of Section 5.1. Let μ' be a probability measure on D'; μ' is defined on the smallest Borel field $\overline{D'}$ which contains all cylinder sets. Let T be the "shift operator" on D'. This means that, if

$$s = (s_1, s_2, \ldots) \varepsilon D',$$

then

$$Ts = (s_1', s_2', \ldots) \varepsilon D',$$

where

$$s_i' = s_{i+1}, \; i = 1, 2, \ldots$$

The probability measure μ' is called stationary if, for any $S \varepsilon \overline{D'}$, $\mu'(T^{-1}S) = \mu'(S)$. A set $S \varepsilon \overline{D'}$ is called invariant if $T^{-1}S = S$. The probability measure μ' is called ergodic if every invariant set has μ'-measure 0 or 1.

Let μ' be a stationary ergodic probability measure. Let

$$X = (X_1, X_2, \ldots)$$

be a sequence of chance variables, such that the range of X is D', and

$$P\{X \varepsilon S\} = \mu'(S)$$

for every $S \varepsilon \overline{D'}$. The chance variable

$$-\log P\{X_{n+1} \mid X_1, \ldots, X_n\} \tag{10.2.1}$$

and its expected value

$$h_{n+1} = H(X_{n+1} \mid X_1, \ldots, X_n) \tag{10.2.2}$$

have been defined in Chapter 2. We shall also be concerned with the chance variable

$$g_n = -\frac{1}{n} \log P(X_1, \ldots, X_n)$$

which, of course, was also defined in Chapter 2. Write $\gamma_n = E g_n$.

From the stationarity of μ' and (2.2.15) we have

$$\begin{aligned} h_{n+2} &= H(X_{n+2} \mid X_1, \ldots, X_{n+1}) \\ &\leq H(X_{n+2} \mid X_2, \ldots, X_{n+1}) = h_{n+1} \end{aligned} \tag{10.2.3}$$

for every n. Hence

$$\lim_{n\to\infty} h_n = h_\infty \text{ (say)}. \tag{10.2.4}$$

Since

$$\gamma_n = \frac{h_1 + \cdots + h_n}{n}, \tag{10.2.5}$$

we have that

$$\lim_{n\to\infty} \gamma_n = h_\infty. \tag{10.2.6}$$

We now give a proof of

Theorem 10.2.1. Let μ' be stationary and ergodic. As $n \to \infty$ the chance variable g_n converges stochastically to h_∞.

McMillan [*1*], to whom the theorem is due, first proved convergence in L^1, which is of course a stronger result (see also Khinchin [*1*]). Recently Breiman [*1*], [*2*] has proved convergence with probability one. The proof which follows is adapted from Thomasian [*1*]. The property which is the conclusion of the theorem will be called the "asymptotic equipartition property" and suffices for all presently known applications to coding theory (see Section 5.6).

Suppose S is a subset of D'_n. We define $M_n(S)$ as the number of n-sequences in S, and $\mu'(S)$ as the μ'-measure of the cylinder set in D' whose base is S.

Theorem 10.2.1 will be an immediate consequence of (10.2.6) and Lemmas 10.2.2 and 10.2.3 to be proved below. Lemma 10.2.1 will be needed for the proof of Lemma 10.2.2.

Lemma 10.2.1. Let $\varepsilon > 0$ be arbitrary. For every sufficiently large n there exists a set $B_n \subset D'_n$ such that

$$M_n(B_n) < 2^{n(h_n+\varepsilon)} \tag{10.2.7}$$

$$\mu'(B_n) > 1 - \varepsilon. \tag{10.2.8}$$

Lemma 10.2.2. Let $\varepsilon > 0$ be arbitrary. For every sufficiently large n we have

$$P\{g_n < \gamma_n + \varepsilon\} > 1 - \varepsilon. \tag{10.2.9}$$

Lemma 10.2.3. Let $\varepsilon > 0$ be arbitrary. For every sufficiently large n we have

$$P\{g_n < \gamma_n - \varepsilon\} < 3\varepsilon \log a. \tag{10.2.10}$$

Proof of Lemma 10.2.1. Because of (10.2.4) it is enough to prove (10.2.8) and the following (10.2.11) ((10.2.11) in place of (10.2.7)): For any fixed positive integer t and all n sufficiently large,

$$M_n(B_n) < 2^{n\left(h_t + \frac{\varepsilon}{2}\right)}. \tag{10.2.11}$$

Let $t > 1$ be any integer. Number the a^t t-sequences in some fixed but arbitrary manner. Let τ_i be the ith t-sequence. Define

$$\begin{aligned}\nu(\tau_i) &= \nu(\tau_{i1}, \ldots, \tau_{it}) \\ &= P\{X_t = \tau_{it} \mid X_1 = \tau_{i1}, \ldots, X_{t-1} = \tau_{i(t-1)}\}\end{aligned} \tag{10.2.12}$$

with the proviso that $\nu(\tau_i) = 0$ if

$$P\{X_1 = \tau_{i1}, \ldots, X_{t-1} = \tau_{i(t-1)}\} = 0.$$

Also define the chance variable

$$N_n(X) = (N_n(X \mid 1), \ldots, N_n(X \mid a^t)) \tag{10.2.13}$$

as follows: For $i = 1, \ldots, a^t$, $N_n(X \mid i)$ is $\frac{1}{n}$ multiplied by the number of k, $k = 1, \ldots, (n - t + 1)$, such that

$$(X_k, X_{k+1}, \ldots, X_{k+t-1}) = \tau_i. \tag{10.2.14}$$

It follows from the stationarity and ergodicity of μ' and the ergodic theorem (Doob [1], p. 464) that $N_n(X)$ converges, with probability one, to a constant vector whose i^{th} component, $i = 1, \ldots, a^t$, is

$$P\{(X_1, \ldots, X_t) = \tau_i\} = \beta(\tau_i), \quad \text{say}. \tag{10.2.15}$$

Let $\varepsilon' > 0$ be arbitrary. It follows that, for every sufficiently large n, there exists a set $B_n \subset D_n'$ such that

$$\mu'(B_n) > 1 - \varepsilon', \tag{10.2.16}$$

and such that any n-sequence $u_0 \varepsilon B_n$ has the following property:

$$|N_n(u_0 \mid i) - \beta(\tau_i)| \leqq \varepsilon' \beta(\tau_i), \quad i = 1, \ldots, a^t. \tag{10.2.17}$$

(The meaning of the symbol $N_n(u_0 \mid i)$ is obvious, but we spell it out to remove any doubt. Let $u_0 = (x_1, \ldots, x_n)$. Then $N_n(u_0 \mid i)$ is $\frac{1}{n}$ multiplied by the number of k, $k = 1, \ldots, (n - t + 1)$, such that

$$(x_k, x_{k+1}, \ldots, x_{k+t-1}) = \tau_i.)$$

We now show that, for sufficiently small ε' and sufficiently large n, (10.2.11) is fulfilled. We do this by applying the argument which was applied repeatedly in Section 2.1. Let

$$Y_1, Y_2, \ldots$$

be a sequence of chance variables, with values in the set $\{1, \ldots, a\}$, and such that, for $n > t$,

$$\begin{aligned}&P\{Y_n = b_n \mid Y_1 = b_1, \ldots, Y_{n-1} = b_{n-1}\} \\ &\quad = P\{X_n = b_n \mid X_{n-t+1} = b_{n-t+1}, \ldots, X_{n-1} = b_{n-1}\}\end{aligned} \tag{10.2.18}$$

for every n-sequence $(b_1, \ldots, b_n)$. (Thus $\{Y_n, n \geqq 1\}$ is a Markov chain of order $(t - 1)$.) The "initial" distribution, which is on $(t - 1)$-sequences, is obviously and uniquely determined by the requirement

$$P\{(Y_1, \ldots, Y_t) = \tau_i\} = \beta(\tau_i), \quad i = 1, \ldots, a^t. \tag{10.2.19}$$

Let $u_0 = (x_1, \ldots, x_n)$ be any n-sequence in B_n. It follows from (10.2.17) that

$$P\{(Y_1, \ldots, Y_n) = u_0\} > \beta(x_1, \ldots, x_t) \cdot \prod_{i=1}^{a^t} \nu(\tau_i)^{n(1+\varepsilon')\beta(\tau_i)} > 0. \tag{10.2.20}$$

Hence

$$M_n(B_n) < d \cdot \exp_2\{n(1+\varepsilon')\, h_t\}, \tag{10.2.21}$$

where d is the reciprocal of the smallest positive $\beta(\tau_i)$. The desired result (10.2.11) follows from (10.2.21) if ε' is taken sufficiently small. This proves Lemma 10.2.1.

Proof of Lemma 10.2.2. Let $\varepsilon > 0$ be arbitrary. For every sufficiently large n let B_n be a set which satisfies (10.2.7) and (10.2.8). Define the following subsets of D_n':

$$G_{4n}' = \left\{u_0 \middle| -\frac{1}{n}\log \mu'(u_0) < \gamma_n + 3\varepsilon\right\}$$

$$G_{5n}' = D_n' - G_{4n}'$$

$$G_{4n} = B_n \cap G_{4n}'$$

$$G_{5n} = B_n \cap G_{5n}'.$$

For $u_0 \varepsilon G_{5n}'$ we have

$$\mu'(u_0) \leqq 2^{-n(\gamma_n+3\varepsilon)}.$$

Then

$$\begin{aligned}\mu'(B_n) &= \mu'(G_{4n}) + \mu'(G_{5n}) \\ &\leqq \mu'(G_{4n}') + M_n(B_n)\, 2^{-n(\gamma_n+3\varepsilon)}.\end{aligned} \tag{10.2.22}$$

From (10.2.4), (10.2.6), and (10.2.7) we conclude that

$$M_n(B_n) < 2^{n(\gamma_n+2\varepsilon)} \tag{10.2.23}$$

for all n sufficiently large. From (10.2.8), (10.2.22), and (10.2.23) we have that, for all n sufficiently large,

$$\mu'(G_{4n}') > 1 - \varepsilon - 2^{-n\varepsilon}. \tag{10.2.24}$$

Replacing ε by $\frac{\varepsilon}{3}$ and taking n sufficiently large we obtain (10.2.9). This proves Lemma 10.2.2.

Proof of Lemma 10.2.3. Let $\varepsilon > 0$ be arbitrary. Define the following subsets of D_n':

$$G_1 = \left\{u_0 \middle| -\frac{1}{n}\log \mu'(u_0) < \gamma_n - \varepsilon\right\}$$

$$G_2 = \left\{u_0 \middle| \gamma_n - \varepsilon \leqq -\frac{1}{n}\log \mu'(u_0) \leqq \gamma_n + \varepsilon^2\right\}$$

$$G_3 = \left\{u_0 \middle| -\frac{1}{n}\log \mu'(u_0) > \gamma_n + \varepsilon^2\right\}.$$

Hence

$$\gamma_n \leqq (\gamma_n - \varepsilon)\,\mu'(G_1) + (\gamma_n + \varepsilon^2)\,(1 - \mu'(G_1)) - \frac{1}{n} \sum_{u_0 \varepsilon G_3} \mu'(u_0) \log \mu'(u_0)$$

$$= (\gamma_n + \varepsilon^2) - (\varepsilon + \varepsilon^2)\,\mu'(G_1) - \frac{1}{n} \sum_{u_0 \varepsilon G_3} \mu'(u_0) \log \mu'(u_0)\,. \qquad (10.2.25)$$

From (10.2.25) we obtain

$$\mu'(G_1) \leqq \varepsilon - \frac{1}{n\,\varepsilon} \sum_{u_0 \varepsilon G_3} \mu'(u_0) \log \mu'(u_0)\,. \qquad (10.2.26)$$

We now show that, for sufficiently large n, the second term in the right member of (10.2.26) is less than $2\varepsilon \log a$. Now G_3 contains at most a^n sequences. Hence, by the argument which led to (2.2.4) we have

$$\begin{aligned} &- \sum_{u_0 \varepsilon G_3} \mu'(u_0) \log \mu'(u_0) \\ &\leqq - \mu'(G_3) \log \mu'(G_3) + n\,\mu'(G_3) \log a\,. \end{aligned} \qquad (10.2.27)$$

From Lemma 10.2.2 we have that, for n sufficiently large,

$$\mu'(G_3) < \varepsilon^2. \qquad (10.2.28)$$

The desired result now follows from (10.2.27) and (10.2.28) when n is made sufficiently large. This proves Lemma 10.2.3.

10.3. Admissibility of an ergodic input for a discrete finite-memory channel. Consider a continuously operating discrete finite-memory channel. Let F, F', and F'' be the spaces defined in Section 6.2, and let $\bar{F}$, $\bar{F}'$, and $\bar{F}''$ be the Borel fields defined on these spaces in Section 6.2. Let Q' be a stationary ergodic probability measure defined on $\bar{F}'$, and let Q be the resulting probability measure on $\bar{F}$ induced by the channel. Q is obviously stationary. We now prove

Theorem 10.3.1. Under the above conditions Q is ergodic.

When Q is ergodic Q' is said to be "admissible". Before proving Theorem 10.3.1 we need to prove several auxiliary lemmas. Our proof is adapted from FEINSTEIN ([*2*], p. 99). The theorem is due to KHINCHIN [*1*].

Lemma 10.3.1. Necessary and sufficient for Q to be ergodic is that, for any two sets A and B in $\bar{F}$, we have

$$\lim_{n\to\infty} \frac{1}{n} \sum_{i=0}^{n-1} Q(T^{-i} A \cap B) = Q(A) \cdot Q(B)\,. \qquad (10.3.1)$$

Proof: Suppose A is such that $TA = A$. Set $B = A$ in (10.3.1). If (10.3.1) holds then $Q(A) = [Q(A)]^2$, i.e., $Q(A) = 0$ or 1. Hence (10.3.1) is sufficient. We now prove the necessity of the condition.

Let f and g be the characteristic functions of the sets A and B, respectively. Suppose Q is ergodic. The ergodic theorem (DOOB [*1*],

p. 464) implies that, for almost every (Q) $u_0 \varepsilon F$,

$$\lim_{n\to\infty}\left[\frac{1}{n}\sum_{i=0}^{n-1} f(T^i u_0)\, g(u_0)\right] = Q(A)\, g(u_0). \qquad (10.3.2)$$

The quantity in square brackets in the left member of (10.3.2) is bounded by 1. Hence, by the Lebesgue bounded convergence theorem,

$$\begin{aligned} &\lim_{n\to\infty}\frac{1}{n}\sum_{i=0}^{n-1}\int_F f(T^i u_0)\, g(u_0)\, Q(du_0) \\ &= \lim_{n\to\infty}\frac{1}{n}\sum_{i=0}^{n-1} Q(T^{-i}A \cap B) = \int_F Q(A)\, g(u_0)\, Q(du_0) \qquad (10.3.3) \\ &= Q(A)\, Q(B). \end{aligned}$$

This proves Lemma 10.3.1.

Lemma 10.3.2. Let A and B be any two cylinder sets (distinct or not) with finite bases. Then (10.3.1) is satisfied for these sets A and B.

Proof: Let a_1 and a_2 be any sequences in D'_n, and b_1 and b_2 be any sequences in D''_n. Let $c_1 = b_1(m+1, n)$, $c_2 = b_2(m+1, n)$, in the notation of Section 5.2. Let $A(B)$ be the set of sequences (u_0, v_0) in F such that the corresponding coordinates of u_0 and v_0 are the same as those of a_1 and c_1 (the same as those of a_2 and c_2). From the stationarity of Q and the form of (10.3.1) it follows that it suffices to show that, for any n, sets A and B of this form satisfy (10.3.1). From the stationarity of Q, (5.2.1), and (5.2.2), it follows that, for i sufficiently large,

$$Q(T^{-i} A \cap B) = Q'(T^{-i} a_1 \cap a_2) \cdot \phi(c_1 \mid a_1) \cdot \phi(c_2 \mid a_2) \quad (10.3.4)$$

where (1) $(T^{-i}a_1 \cap a_2)$ is the cylinder set in F' of all sequences whose coordinates with indices $1, \ldots, n$ are the same as those of a_2 and whose coordinates with indices $(1+i), (2+i), \ldots, (n+i)$ are the same as those of a_1 (2) ϕ is an appropriate function of the arguments exhibited.

Now, in (10.3.1), replace the expression in the left member of (10.3.4) by the expression in the right member. Recall that Q' is ergodic, and apply the necessity condition of Lemma 10.3.1 to the sequence which results from dividing by $\phi(c_1 \mid a_1) \cdot \phi(c_2 \mid a_2)$. Lemma 10.3.2 follows.

Let B now be any fixed cylinder set with a finite base. Let K be the class of all sets in $\overline{F}$ such that, if $A \varepsilon K$, A and (the now fixed) B satisfy (10.3.1). We now prove

Lemma 10.3.3. If $A_1, A_2, \ldots$ is a monotonic sequence of sets in K, then $A_\infty = \lim_i A_i$ is also in K.

Proof of Lemma 10.3.3. Let $A_1, A_2, \ldots$ be a monotonically non-decreasing sequence of sets in K. From the definition of K we have

$$\lim_{n\to\infty} \frac{1}{n} \sum_{i=0}^{n-1} Q(T^{-i} A_k \cap B) = Q(A_k) \cdot Q(B) \tag{10.3.5}$$

for any integer k. Hence

$$\liminf_{n\to\infty} \frac{1}{n} \sum_{i=0}^{n-1} Q(T^{-i} A_\infty \cap B) \geqq Q(A_k) \cdot Q(B). \tag{10.3.6}$$

Let $k \to \infty$ in (10.3.6). We obtain

$$\liminf_{n\to\infty} \frac{1}{n} \sum_{i=0}^{n-1} Q(T^{-i} A_\infty \cap B) \geqq Q(A_\infty) \cdot Q(B). \tag{10.3.7}$$

Returning to (10.3.5), we have, for any integer k,

$$\lim_{n\to\infty} \frac{1}{n} \sum_{i=0}^{n-1} Q(T^{-i} A_k \cap B) \leqq Q(A_\infty) \cdot Q(B). \tag{10.3.8}$$

Since, by the stationarity of Q,

$$Q(T^{-i} A_\infty \cap B) - Q(T^{-i} A_k \cap B) \leqq Q(A_\infty - A_k), \tag{10.3.9}$$

and since the right member of (10.3.9) approaches zero as $k \to \infty$, we have, from (10.3.8),

$$\limsup_{n\to\infty} \frac{1}{n} \sum_{i=0}^{n-1} Q(T^{-i} A_\infty \cap B) \leqq Q(A_\infty) \cdot Q(B). \tag{10.3.10}$$

From (10.3.7) and (10.3.10) we obtain that A_∞ is in K.

To prove the lemma for a monotonically non-increasing sequence of sets A_i, apply the result of the preceding paragraph to the monotonically non-decreasing sequence of their complements $\bar{A}_i$. This proves Lemma 10.3.3.

Proof of Theorem 10.3.1. It follows from Lemma 10.3.2 that K contains the finitely-additive field of all finite unions of cylinder sets with finite bases. Hence it follows from Lemma 10.3.3 that $K = \bar{F}$ for *any* cylinder set B with a finite base.

Now let A be any fixed set in $\bar{F}$, and let L be the class of all sets in $\bar{F}$ such that, if $B \varepsilon L$, (the now fixed) A and B satisfy (10.3.1). Then an argument almost the same as that of Lemma 10.3.3 (which made no use of the fact that B was a cylinder set) has as a consequence that $L = \bar{F}$. The theorem now follows from Lemma 10.3.1.

11. Fundamentals of Rate Distortion Theory

11.1. Introduction. Let $X^1, \ldots, X^n$ be independent chance variables, each with distribution $\pi = (\pi_1, \ldots, \pi_a)$ on the set $A = \{1, \ldots, a\}$ (the "source" alphabet). Let $X_n = (X^1, \ldots, X^n)$, and A_n^* be the space of all n-sequences with elements in the source alphabet. Let $C = \{1, \ldots, c\}$ be the "reproduction" alphabet, and C_n^* the space of all n-sequences with elements in this alphabet. Let d be a given "distortion" function, i.e., $d(i, j)$ is the "distortion" between i $(i = 1, \ldots, a)$ and j $(j = 1, \ldots, c)$. Define $d_0' = \min_{i,j} d(i, j)$, $d_{00}' = \max_{i,j} d(i, j)$. Let $a_n = (a^1, \ldots, a^n) \varepsilon A_n^*$ and $c_n = (c^1, \ldots, c^n) \varepsilon C_n^*$. The distortion $d(a_n, c_n)$ between a_n and c_n is defined to be

$$d(a_n, c_n) = \frac{1}{n} \sum_{j=1}^{n} d(a^i, c^i) . \tag{11.1.1}$$

Let z be a parameter which temporarily will take values in (d_0', d_{00}'). For any such z let $w(\cdot|\cdot|z)$ be a c.p.f., such that $w(j|i|z)$ for $i = 1, \ldots, a$; $j = 1, \ldots, c$, and

$$\sum_{i,j} \pi_i w(j|i|z) d(i, j) = L(w(\cdot|\cdot|z)) \text{ (say)} \leqq z \tag{11.1.2}$$

and, defining $R(\pi, w(\cdot|\cdot|z))$ as in Chapter 6,

$R(\pi, w(\cdot|\cdot|z))$ is a minimum for all c.p.f.'s which satisfy (11.1.2). (11.1.3)

(The minimum in (11.1.3) obviously exists.) The rate distortion function D is defined by

$$D(z) = R(\pi, w(\cdot|\cdot|z)) . \tag{11.1.4}$$

The significance of D will now be explained. We want to "span" the space A_n^*, with its accompanying probability distribution implied by the distribution of X_n, to within average distortion z, by a set $C_n(z)$ of sequences in C_n^* in the following way: For each element $a_n \in A_n^*$ let $\gamma(a_n)$ be an element of $C_n(z)$ (it need not be unique) such that

$$d(a_n, \gamma(a_n)) = \min_{c_n \in C_n(z)} d(a_n, c_n) . \tag{11.1.5}$$

We desire that

$$\mathrm{E} d(X_n, \gamma(X_n)) \leqq z \tag{11.1.6}$$

and

$$\left\{ \begin{array}{l} \text{the number of sequences in } C_n(z) \\ \text{to be essentially minimal} \end{array} \right. . \tag{11.1.7}$$

We will see that the function D plays a crucial role in the solution of this problem.

The largest value d_{00} of z which we need to consider is the smallest value of z obtainable when $C_n(z)$ consists of a single sequence. A best such sequence is one everyone of whose elements is j_{00}, where j_{00} is any j such that

$$\sum_i \pi_i d(i, j_{00}) = \min_j \sum_i \pi_i d(i, j) = d_{00}\,. \tag{11.1.8}$$

The minimum value d_0 of z which we can achieve is obviously

$$d_0 = \Sigma \pi_i \min_j d(i, j)\,. \tag{11.1.9}$$

From now on z will be assumed to lie in (d_0, d_{00}).

In this interval the function D is obviously monotonically non-increasing. The inequality sign in (11.1.2) and (11.1.6) can obviously be replaced by an equality sign. Let us now prove that D is convex. (Hence D is continuous and strictly decreasing.) Suppose $z_1 < z_2$ are points in the interval (d_0, d_{00}). Let $w(\cdot|\cdot|z_1)$ and $w(\cdot|\cdot|z_2)$ be corresponding $w(\cdot|\cdot|\cdot)$. Now define (just for this argument) $w_0(\cdot|\cdot) = \frac{1}{2} w(\cdot|\cdot|z_1) + \frac{1}{2} w(\cdot|\cdot|z_2)$. Then $L(w_0) = \frac{1}{2} L(w(\cdot|\cdot|z_1)) + \frac{1}{2} L(w(\cdot|\cdot|z_2))$. We already know (argument preceding Lemma 6.4.2) that, for fixed π, $R(\pi, w)$ is a convex function of w. Then we have

$$R\left(\pi, w\left(\cdot|\cdot|\frac{z_1+z_2}{2}\right)\right) \le R(\pi, w_0)$$

$$\le \frac{1}{2} R(\pi, w(\cdot|\cdot|z_1)) + \frac{1}{2} R(\pi, w(\cdot|\cdot|z_2))\,. \tag{11.1.10}$$

11.2. The approximation theorem. The positive part of the result we desire is contained in the following.

Theorem 11.2.1. Let $\varepsilon > 0$ be arbitrary. For n (depending on ε) sufficiently large, there exists, for any z in (d_0, d_{00}), a set $C_n(z)$ such that

$$\mathrm{E} d(X_n, \gamma(X_n)) < z + \varepsilon\,, \tag{11.2.1}$$

and the number $N(z)$ of elements in $C_n(z)$ satisfies

$$N(z) \leqq \exp_2\{nD(z)\}\,. \tag{11.2.2}$$

First we need a simple lemma which we will also need in a later chapter.

Lemma 11.2.1. Consider the code construction of Theorem 3.2.1. Let $\varepsilon' > 0$ be arbitrary. By making both the δ of (2.1.3) and $\lambda < 1$ sufficiently large, we can rewrite (3.2.6) as

$$P\{Y \varepsilon A_1 \cup \cdots \cup A_N\} > 1 - \varepsilon'\,. \tag{11.2.3}$$

Proof: This follows from (3.2.6) after we take note of the following facts: a) The $\frac{\lambda}{2}$ of (3.2.6) was arbitrary. It could have been replaced by any positive number less than λ. (See (3.2.5)) b) By making δ sufficiently large we can replace $\frac{3}{4}$ by any number less than one.

We now proceed to the proof of Theorem 11.2.1. Let $(Z^1, \ldots, Z^n) = Z_n$ be n independent chance variables, each with the same distribution π' implied by

$$P\{X^1 = i, Z^1 = j\} = \pi_i w(j|i|z) . \tag{11.2.4}$$

The "reverse" c.p.f. $w'(\cdot|\cdot|z)$ is to be defined by

$$\pi_i w(j|i|z) = \pi'_j w'(i|j|z) . \tag{11.2.5}$$

With the δ of (2.1.3) and λ, $0 < \lambda < 1$, temporarily fixed, we construct, by the method of Theorem 3.2.1, a code whose transmitted sequences are elements of C^*_n, received sequences are elements of A^*_n, and c.p.f. is $w'(\cdot|\cdot|z)$. Let $(u_1, A_1), \ldots, (u_N, A_N)$ be the elements of this code. We now show that, if we write

$$C^0_n(z) = \{u_1, \ldots, u_N\} , \tag{11.2.6}$$

this set will "essentially" have the properties required of $C_n(z)$ when δ and $\lambda < 1$ are large enough. Then we will remove the qualifying word "essentially".

Let $d_{\max} = \max_{i,j} d(i, j)$. When δ and λ are sufficiently large, we have from Lemma 11.2.1 that

$$P\{X_n \varepsilon A_1 \cup \cdots \cup A_N\} > 1 - \frac{\varepsilon}{4 d_{\max}} \cdot \tag{11.2.7}$$

Hence the contribution of $A^*_n - \bigcup_{i=1}^{N} A_i$ to the left member of (11.2.1) (when the range of γ is in $C^0_n(z)$) is less than $\frac{\varepsilon}{4} \cdot$ Since every n-sequence in A_i is generated by its code word, it follows, from the definition of generation and (11.1.1), that, when n is sufficiently large, the distortion between any member of A_i and its transmitted word u_i is less that $\frac{\varepsilon}{4} + z$. Hence (11.2.1) is satisfied with ε replaced by $\frac{\varepsilon}{2} \cdot$ From Lemma 3.3.1 we have, since

$$D(z) = R(\pi, w(\cdot|\cdot|z)) = R(\pi', w'(\cdot|\cdot|z)) , \tag{11.2.8}$$

that, when n is sufficiently large,

$$N < \exp_2 \{n [D(z) + \varepsilon'']\} . \tag{11.2.9}$$

The qualification "essentially" was made because of the presence of ε'' in the right member of (11.2.9). A continuity argument will serve to remove it. In the construction of the code use $w(\cdot|\cdot|z+\eta)$ for $\eta > 0$. If η is sufficiently small and n sufficiently large, we can still have (11.2.1) satisfied. Since D is strictly decreasing, we can, by making n large enough, replace (11.2.9) by (11.2.2). This completes the proof of Theorem 11.2.1.

11.3. Converse of the approximation theorem. A natural question which arises is whether the right member of (11.2.2) can be reduced. In this section we will prove that this cannot be done, modulo our usual change of sign of ε in the exponent. In the present case this will mean replacing the right member of (11.2.1) by $\leqq z$. (See Theorem 11.3.3 below.)

Let X and Y be chance variables defined on the same probability space. For simplicity of exposition we will assume that their ranges are finite. We define

$$\bar{I}(X, Y) = H(X) - H(X \mid Y) = H(Y) - H(Y \mid X) .$$

We have already encountered this function, "the mutual information between X and Y", in many guises, e.g., (2.2.1), $R(\pi, w)$ in Chapter 6, and throughout Chapter 7. The following theorem is known in the literature as the *data processing* theorem.

Theorem 11.3.1. Let the chance variables Z_1, Z_2, Z_3, Z_4, which take only finitely many values, form a Markov chain (in this order). Then

$$\bar{I}(Z_1, Z_4) \leqq \bar{I}(Z_2, Z_3) . \tag{11.3.1}$$

The justification of the name is this: $\bar{I}$ is thought of as "the mutual information". Suppose the chance variable Z_1 is processed (transformed) into the chance variable Z_2, the latter is "transmitted over a channel", received as Z_3, and then processed into Z_4. The theorem says that no "information" can be gained thereby. The description of the theorem is suggestive and helpful if not pushed too far. The theorem could be thought of as a sort of law of conservation of information. The proof is very simple. We have

$$\begin{aligned} \bar{I}(Z_2, Z_3) &= H(Z_3) - H(Z_3|Z_2) \\ &= H(Z_3) - H(Z_3|Z_1, Z_2) \quad \text{by the Markov property} \\ &\geqq H(Z_3) - H(Z_3|Z_1) \quad \text{by} \quad (2.2.15) \\ &= \bar{I}(Z_1, Z_3) . \end{aligned}$$

Thus

$$\bar{I}(Z_1, Z_3) \leqq \bar{I}(Z_2, Z_3) . \qquad (11.3.2)$$

Since Z_1, Z_2, Z_4 is also a Markov chain, we have, from (11.3.2), that

$$\bar{I}(Z_1, Z_4) \leqq \bar{I}(Z_2, Z_4) . \qquad (11.3.3)$$

Since Z_4, Z_3, Z_2 constitute, in this order, a Markov chain, and since $\bar{I}$ is symmetric in its two chance variables, we obtain from (11.3.2) that

$$\bar{I}(Z_2, Z_4) \leqq \bar{I}(Z_2, Z_3) . \qquad (11.3.4)$$

The desired result now follows from (11.3.3) and (11.3.4).

Now let X_n and its distribution be as defined in Section 11.1. Let $Z_n = (Z^1, \ldots, Z^n)$ be a chance variable defined on C_n^*. The chance variables Z^i need not be independent or identically distributed, and the joint distribution of X_n and Z_n may be any distribution consistent with the already specified distribution of X_n. We also need the following

Theorem 11.3.2. Define $\bar{d}$ by

$$n\bar{d} = \sum_{k=1}^{n} \mathrm{E}d(X^k, Z^k) . \qquad (11.3.5)$$

Then

$$\bar{I}(X_n, Z_n) \geqq nD(\bar{d}) . \qquad (11.3.6)$$

Proof: Since the X^k are independent,

$$H(X_n) = \sum_{k=1}^{n} H(X^k) . \qquad (11.3.7)$$

Also

$$H(X_n | Z_n) = \sum_{k=1}^{n} H(X^k | Z_n, X^1, \ldots, X^{k-1}) \leqq \sum_{k=1}^{n} H(X^k | Z^k) \qquad (11.3.8)$$

by repeated application of (2.2.2) and (2.2.15). Hence

$$\bar{I}(X_n, Z_n) = H(X_n) - H(X_n | Z_n)$$

$$\geqq \sum_{k=1}^{n} \bar{I}(X^k, Z^k) \geqq \sum_{k=1}^{n} D(\mathrm{E}d(X^k, Z^k)) , \qquad (11.3.9)$$

by (11.3.7), (11.3.8), and the definition of D. Since D is convex, the last member of (11.3.9) is not less than

$$nD\left(\frac{1}{n} \sum_{k=1}^{n} \mathrm{E}d(X^k, Z^k)\right) = nD(\bar{d}) .$$

We are now ready to prove

Theorem 11.3.3. If

$$\mathrm{E}\, d(X_n, \gamma(X_n)) \leqq z\,, \tag{11.3.10}$$

then

$$N(z) \geqq \exp_2\{nD(z)\}\,. \tag{11.3.11}$$

Proof: On the space C_n^* we induce a probability distribution by

$$P\{Z_n = \gamma(x_n) \mid X_n = x_n\} = 1\,.$$

(For this purpose $\gamma(\cdot)$ is to be uniquely defined.) By Theorem 11.3.2 and (11.3.10), since $z \geqq \bar{d}$ and D is monotonically decreasing,

$$\bar{I}(X_n, Z_n) \geqq nD(z)\,. \tag{11.3.12}$$

On the other hand,

$$\bar{I}(X_n, Z_n) \leqq H(Z_n) \leqq \log N(z)\,. \tag{11.3.13}$$

The theorem follows from (11.3.12) and (11.3.13).

Another proof of Theorem 11.3.3 could be given along the lines of the proof of Theorem 11.2.1. First one separates the members of $C_n(z)$ according to their composition, as in the proof of Theorem 3.3.1. Now consider any one set of elements with the same composition. For each element z_n in this set, and every sequence x_n, generated by z_n with respect to the appropriate $w'(\cdot|\cdot|z)$ of (11.2.5), the distortion $d(x_n, z_n)$ has a proper value. As in the proof of Lemma 11.2.1 and Theorem 11.2.1, the code constructed by the method of Theorem 3.2.1 will have to be sufficiently long in order for the union of its decoding sets to have probability close to 1. Repeating this argument over all the sets of the decomposition, we find a longest code whose length will satisfy (11.3.11) and whose code words must be in $C_n(z)$. We shall not give the details of this proof.

11.4. Summary of the previous results. What has been proved in Theorems 11.2.1 and 11.3.3 is equivalent to the following: Let $C_n(z)$ now be a set whose number of sequences is minimal among all sets for which (11.2.1) holds. Then

$$D(z) = \lim_{\varepsilon \downarrow 0} \lim_{n \to \infty} \frac{1}{n} \log \left(\text{number of sequences in } C_n(z)\right) = \min \bar{I}(X, Z),$$

where X and Z are chance variables defined on the same probability space, X has the distribution π, Z takes values in the set $\{1, \ldots, c\}$, and the minimum is with respect to all such chance variables Z which satisfy

$$\mathrm{E}d(X, Z \leq)\, z\,.$$

11.5. The rate distortion function when side information is available. Consider the following diagram:

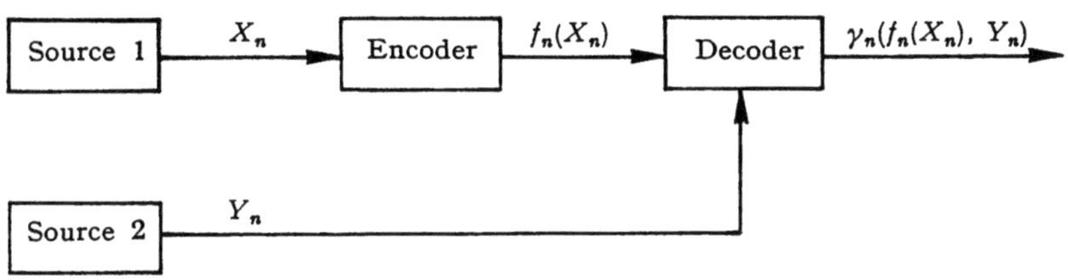

We adopt the notation of Section 11.1 and the following additional notation: The chance variables (X^i, Y^i), $i = 1,\ldots, n$, are independently and identically distributed. $Y_n = (Y^1,\ldots, Y^n)$. Let (X, Y) be a chance variable with the same distribution as (X^i, Y^i). Y takes values in the set $B = \{1,\ldots, b\}$. B_n^* is, as usual, the Cartesian product of $n\,B$'s. The "encoding" function $f_n(X_n)$ takes values in the set

$$\{1, 2,\ldots, \|f_n\|\}\,, \tag{11.5.1}$$

i.e., it numbers the sequences of A_n^*, but not in a one-to-one manner. The "decoding" function $\gamma_n(f_n(X_n), Y_n)$ takes values in C_n^*, the space of n-sequences in the reproduction alphabet. Our problem is to find the rate distortion function $D(\cdot\,|\,\mathrm{dec})$ of X, when the decoder knows the value of Y, i.e., Y_n is one of the arguments of γ_n. We have

$$D(z\,|\,\mathrm{dec}) = \lim_{\varepsilon \downarrow 0}\ \lim_{n\to\infty} \frac{1}{n}\log\,(\text{minimal } \|f_n\|) \tag{11.5.2}$$

where f_n and γ_n satisfy

$$\mathrm{E}d\,(X_n, \gamma_n(f_n(X_n), \gamma_n) < z + \varepsilon\,. \tag{11.5.3}$$

The intuitive meaning of the problem is the same as in Section 11.1. The set (11.5.1) plays the role of counting the number of elements in $C_n(z)$, so that we want $\|f_n\|$ to be minimal. An actual problem might arise in the following way: Suppose X^i is the temperature in Urbana, and Y^i the temperature in Chicago, 150 miles away. It is desired to transmit the temperature in Urbana to the central point in Chicago, while tolerating an average long term distortion per temperature reading of a little more than z. Then $\exp_2\{nD(z|\text{dec})\}$ is the approximate number of sequences needed for n (large) readings. Thus, if a binary alphabet, say, is being used for transmission, approximately $nD(z|\text{dec})$ digits will be needed to transmit the temperature readings for n days.

The function $D(\cdot|\text{dec})$ is given by the following:

Theorem 11.5.1. Let Z be a chance variable with values in the set $S = \{1, \ldots, (a+1)\}$, such that Y, X, Z, in this order, form a Markov chain, and such that there exists a function $\psi: B \times S \to C$, such that

$$\mathrm{E}\, d(X, \psi(Y, Z)) \leqq z\,. \tag{11.5.4}$$

Then

$$D(z|\text{dec}) = \min\, [\bar{I}(X, Z) - \bar{I}(Y, Z)]\,, \tag{11.5.5}$$

where the minimum is with respect to all chance variables Z which satisfy the above conditions.

The proof of the theorem is postponed to Chapter 13, because an essential tool from Chapter 12 will be needed. The function $D(\cdot|\text{dec})$ is computable to any desired degree of accuracy.

Consider now a similar problem, where Y_n is known *only* to the encoder but not to the decoder. (The necessary alteration in the diagram has the Y_n arrow ending at the encoder, but not at the decoder.) Call the rate distortion function of X for this problem, $D(\cdot|\text{enc})$. It is clear that the encoder cannot consolidate any of the contours of the function γ of Sections 11.1 and 11.2, because the decoder has nothing with which to locate X_n further. We therefore conclude:

$$D(z) \overset{z}{\equiv} D(z|\text{enc})\,. \tag{11.5.6}$$

Finally, consider the problem where *both* the encoder and the decoder known Y_n. Now call $D(\cdot|\text{both})$ the corresponding rate distortion function of X. The diagram has the Y_n arrow ending both at the encoder and at the decoder.

When $Y = i$ let $\mu_i(\cdot)$ be the distribution of X. Write $D^i(\cdot)$ for the rate distortion function of X when the latter has distribution $\mu_i(\cdot)$. Let $\pi'_i = P\{Y = i\}$. It is obvious that

$$D(z|\text{both}) = \min \sum_{i=1}^{b} \pi'_i D^i(z_i)\,, \tag{11.5.7}$$

where the minimum is with respect to all $(z_1, \ldots, z_b)$ such that

$$\sum_{i=1}^{b} \pi_i' z_i = z . \tag{11.5.8}$$

An easy argument with a Lagrange multiplier shows the following: If $(z_1^0, \ldots, z_b^0)$ is a minimizing set, then

$$\left. \frac{dD^i(z)}{dz} \right|_{z = z_i^0} \tag{11.5.9}$$

is the same for all i. (It is assumed that all $\pi_i' > 0$.)

We will now give another representation of $D(\cdot|\text{both})$. Let Z_0 be a chance variable with values in C, defined on the same sample space as (X, Y). Suppose that, when $Y = i$, the conditional distribution of (X, Z_0) is such that

$$\mathrm{E}d\,(X, Z_0 | Y = i) = z_i^0 , \tag{11.5.10}$$

and

$$D^i(z_i^0) = \bar{I}\,(X, Z_0 | Y = i) . \tag{11.5.11}$$

Then

$$\mathrm{E}d(X, Z_0) = z \tag{11.5.12}$$

and

$$D\,(z|\text{both}) = \bar{I}(X, Z_0 | Y) . \tag{11.5.13}$$

We conclude that, if Z is a chance variable with values in C, defined on the same sample space as (X, Y), and such that $\mathrm{E}d(X, Z) = z$, then

$$D\,(z|\text{both}) = \min \bar{I}(X, Z | Y) , \tag{11.5.14}$$

where the minimum is with respect to all chance variables Z which fulfill the above requirements.

Remarks

The introduction of the distortion function and Theorem 11.2.1 are due to SHANNON [*5*]. The present proof is due to WOLFOWITZ [*8*]. The idea of this proof simplifies the proofs of Theorems 12.2.1, 12.3.1, 13.1.1., and 13.2.1. The short proof of Theorem 11.3.3 is due to GALLAGER [*1*], p. 447–448. The much longer proof sketched after Theorem 11.3.3 was given earlier by WOLFOWITZ [*8*]. Theorem 11.5.1 is due to WYNER and ZIV [*1*].

12. Source Coding

12.1. Separate coding to span the product of two spaces. Let $(X^1, Y^1), \ldots, (X^n, Y^n)$ be n independent chance variables with the common distribution given by

$$P\{X^1 = i\,,\ Y^1 = j\} = \pi_i w(j|i)\,,\ i = 1, \ldots, a; j = 1, \ldots, b\,. \quad (12.1.1)$$

As usual, the distribution of X^1 will be written as π, that of Y^1 as π'. We will write their joint distribution for short as πw and also as $\pi' w'$, where the definition of w' is obvious (see also (11.2.5)). Define $X_n = (X^1, \ldots, X^n)$ and $Y_n = (Y^1, \ldots, Y^n)$. The spaces on which they are defined will be denoted by A_n^* and B_n^*, respectively, as in Chapter 11. The statement that a certain subset of a space of sequences, on which a given probability distribution has been defined, "covers" the space, is to mean that the probability of the subset can be made $> 1 - \lambda$, with $\lambda > 0$ arbitrarily small.

Returning to Section 2.1 for a moment, we will now define π-sequences and generated sequences more generously than is done there. In (2.1.1) a larger coefficient will be used in the right member, and in (2.1.3) δ will always be chosen sufficiently large (both depend on λ). It then follows, just as in Lemmas 2.1.1 and 2.1.7, that

A_n^* can be covered by at most
$\exp_2\{nH(\pi) + \sqrt{n}K(\lambda)\}$ π-sequences
and cannot be covered by fewer (12.1.2)
than $\exp_2\{nH(\pi) - \sqrt{n}K(\lambda)\}$ sequences

$(A_n^* \times B_n^*)$ can be covered by at
most $\exp_2\{nH(\pi w) + \sqrt{n}K(\lambda)\}$
πw-sequences and cannot be covered (12.1.3)
by fewer than $\exp_2\{nH(\pi w) - \sqrt{n}K(\lambda)\}$
πw-sequences.

The symbol $K(\lambda)$ is a generic designation and need not be the same whenever it occurs.

Thus an "encoder" could number (encode) certain sequences (the πw-sequences) of $(A_n^* \times B_n^*)$ from 1 to at most $\exp_2\{nH(\pi w + \sqrt{n}K(\lambda)\}$, and these sequences would cover $A_n^* \times B_n^*$. The other sequences could be ignored or arbitrarily assigned some number, e.g., 1. The encoding is done by a function, say f, from $A_n^* \times B_n^*$ to the numbers

$$1, \ldots, \exp_2\{nH(\pi w) + \sqrt{n}K(\lambda)\}\,.$$

Suppose now that this encoder is split into two non-communicating encoders. The first of the latter encodes only sequences in A_n^* from 1 to a^*, say, by means of the function f_1, and the second of the latter encodes only sequences in B_n^* from 1 to b^*, say, by means of the function f_2. Does there exist a "decoder", i.e., a decoding function f from the $a^* b^*$ pairs of numbers to sequences in $A_n^* \times B_n^*$, such that

$$f[f_1(x_n), f_2(y_n)] = (x_n, y_n) \tag{12.1.4}$$

for all (x_n, y_n) in a subset of $A_n^* \times B_n^*$ which covers the latter? (The question should really have been phrased thus: Do there exist encoders f_1 and f_2 and a decoder f such that (12.1.4) holds for a subset which covers $A_n^* \times B_n^*$?) Obviously we must have

$$a^* b^* > \exp_2\{n H(\pi w) - \sqrt{n} K(\lambda)\}.$$

For, if this were not so, the encoder who encodes the sequences in $A_n^* \times B_n^*$ could act as each of the separate encoders, in violation of the second part of (12.1.3). The non-trivial question therefore is, whether there exist encoding schemes (functions) for the two encoders and a decoding scheme for the decoder such that

$$a^* b^* < \exp_2\{n H(\pi w) + \sqrt{n} K(\lambda)\}. \tag{12.1.5}$$

The suprising answer is in the affirmative. There exist such encoding and decoding schemes with

$$a^* = \exp_2\left\{n H(\pi) + \frac{\sqrt{n}}{2} K(\lambda)\right\},$$
$$b^* = \exp_2\left\{n \left[\sum_{i=1}^{a} \pi_i H(w(\cdot|i))\right] + \frac{\sqrt{n}}{2} K(\lambda)\right\}, \tag{12.1.6}$$

or the pair obtained by reversing the roles of X^1 and Y^1. (Recall that $K(\lambda)$ need not be the same in all expressions. The 2 which divides $\sqrt{n}$ in (12.1.6) is therefore unnecessary. It will be convenient later.)

Theorem 12.1.1. The pair (12.1.6) (or the pair obtained by reversing the roles of X^1 and Y^1) can be achieved, i.e., the corresponding encoding and decoding schemes exist.

Proof: The encoder who numbers the sequences in A_n^* will simply number the π-sequences seriatim until their probabilities add to $1 - \frac{\lambda}{4}$, at least. The crux of the matter will be the numbering of B_n^* by the second encoder.

Each π-sequence u_0, say, in A_n^*, generates the set $G(u_0)$, say, of generated sequences, whose number is at most the b^* of (12.1.6) and whose conditional probability is at least $1 - \frac{\lambda}{4}\cdot$ (Lemma 2.1.6). If all the $G(u_0)$ were disjoint the problem would be trivial, since then one could number the sequences in each $G(u_0)$ arbitrarily from 1 to b^*. Every sequence generated by some π-sequence would carry such a number, and the totality of all pairs, each consisting of a π-sequence and a sequence generated by it, covers $(A_n^* \times B_n^*)$, by the argument of Lemmas 2.1.1 and 2.1.2 with our present (new) coefficients. Given a pair of numbers from the encoder, the first number would uniquely identify the π-sequence, and consequently the second would uniquely identify the sequence generated by the π-sequence. However, the various $G(u_0)$ are not disjoint, but overlap considerably. If the reader has any doubt about the matter, the following argument will dispel it. If they were disjoint, the total number of sequences generated by all π-sequences would be $a^* b^*$, which would violate Lemma 2.1.4 for large n.

In numbering the sequences in B_n^* which are generated by the π-sequences of A_n^* we are therefore confronted with the following difficulty: Since b^* is much smaller than the total number of sequences in B_n^* which we want to number, many of the latter will have to bear the same number. (This number is approximately, i.e., to within $\sqrt{n}K(\lambda)$ in the exponent,

$$\exp_2\left\{n\left[H(\pi') - \sum_{i=1}^{a} \pi_i H(w(\cdot\,|\,i))\right]\right\},$$

by Lemmas 2.1.4 and 2.1.6.) Yet, within *every* $G(u_0)$, whose number is approximately

$$\exp_2\{nH(\pi)\},$$

every sequence must have a *different* number. This is so because, if we are to be able to identify the pairs (x_n, y_n) which cover $(A_n^* \times B_n^*)$, we certainly must be able to identify the member y_n.

After this heuristic description of the heart of the problem, we proceed to the very simple precise argument. Let each $G(u_0)$ contain at most

$$\exp_2\left\{n\left[\sum_{i=1}^{a} \pi_i H(w(\cdot\,|\,i))\right] + \frac{\sqrt{n}}{4} K(\lambda))\right\} \tag{12.1.7}$$

sequences. To every sequence in $\bigcup G(u_0)$ we assign at random (i.e., with equal probability) one of the numbers

$$1,2,\ldots, \exp_2\left\{n\left[\sum_{i=1}^{a} \pi_i H(w(\cdot\,|\,i))\right] + \frac{\sqrt{n}}{2} K(\lambda)\right\}. \tag{12.1.8}$$

An error in decoding an encoding procedure occurs if, and only if, two sequences in some $G(u_0)$ have the same number. The expected value of the probability of an error in decoding the random encoding is therefore less than

$$\begin{aligned}
&\sum_{u_0} \sum_{y_n \varepsilon G(u_0)} P\{X_n = u_0, Y_n = y_n\} \\
&\cdot (\text{number of sequences in } G(u_0)) \\
&\cdot \left(\exp_2\left\{-n\left[\sum_{i=1}^{a} \pi_i H(w(\cdot|i))\right] - \frac{\sqrt{n}}{2} K(\lambda)\right\}\right) \qquad (12.1.9) \\
&\leqq \exp_2\left\{-\frac{\sqrt{n}}{4} K(\lambda)\right\} \sum_{u_0} \sum_{y_n \varepsilon G(u_0)} P\{X_n = u_0, Y_n = y_n\} \\
&< \exp_2\left\{-\frac{\sqrt{n}}{4} K(\lambda)\right\} < \frac{\lambda}{2}
\end{aligned}$$

when n is sufficiently large, by (12.1.7). Thus, when n is large, there exists a (deterministic) encoding with probability of decoding error $< \frac{\lambda}{2}$. The total error of the procedure is therefore $< \lambda$. This proves the theorem.

Suppose f_1 takes values from 1 to M_1, and f_2 takes values from 1 to M_2. Then

$$\left(\frac{1}{n} \log M_1, \frac{1}{n} \log M_2\right)$$

is called the rate or the rate pair of the encoding procedure (f_1, f_2). A point (or pair) (r_1°, r_2°) is called "achievable" or "attainable" if, for any λ and $\varepsilon > 0$ and n sufficiently large, there exist f_1, f_2 with rate pair

$$(r_1^\circ + \varepsilon, r_2^\circ + \varepsilon) .$$

Thus Theorem 12.1.1 has the following consequence: Let (X, Y) have the same distribution as (X^1, Y^1). Then the points

$$r_1 = (H(X), H(Y|X))$$

and

$$r_2 = (H(X|Y), H(Y))$$

are achievable. Actually we shall see that any point on the line joining r_1 and r_2 is achievable. Suppose first that a point on the line divides the line in the ratio of two integers, m_1 and m_2. Then one uses the scheme corresponding to the proper one of the end points r_1, r_2, for a fraction $\frac{m_1}{m_1 + m_2}$ of the n chance variables (of the "time"), and the scheme corresponding to the other end point for a fraction $\frac{m_2}{m_1 + m_2}$ of the n chance variables (of the time). This procedure is called "time sharing". Any point which is the limit of achievable points is obviously itself achievable.

The above argument need not be limited to points on the line joining r_1 and r_2, and shows that the set of achievable points is convex and closed. Any point, each of whose coordinates is $\geqq$ the corresponding coordinates of an achievable point, is obviously achievable, by "time wasting". Let us call a set M-closed (resp., m-closed) if, with any point r in it, it contains all points whose coordinates are $\geqq$ (resp., $\leqq$) the corresponding coordinates of r. Thus the set of achievable points in our problem is M-closed. Let S be any set. The smallest closed, convex, M-closed (resp., m-closed) set which contains S will be called the M-closure (resp., m-closure) of S.

Another convention to which we shall adhere throughout the remainder of the monograph is this: Suppose f_0 is an encoding function. Then it will always take values in the set $\{1, 2, \ldots, \|f_0\|\}$, i.e., $\|f_0\|$ is the maximum value it will take.

Finally, before proceeding with the argument, we wish to make one more remark which will apply for the remainder of the monograph. We have already said, before (12.1.2), that, on that particular occasion, the multiplicative constants in the definition of π-sequence and of generation were to be made larger than those given in Chapter 2. In future, in this chapter and others following, we shall use arguments which require the multiplicative constants to be suitably chosen. The choice will always be obvious and will not be specifically referred to. In particular, we will no longer be limited to the constants of Chapter 2.

We now introduce the reader to the suggestive graphs employed by engineers. The problem of this section is described graphically as follows:

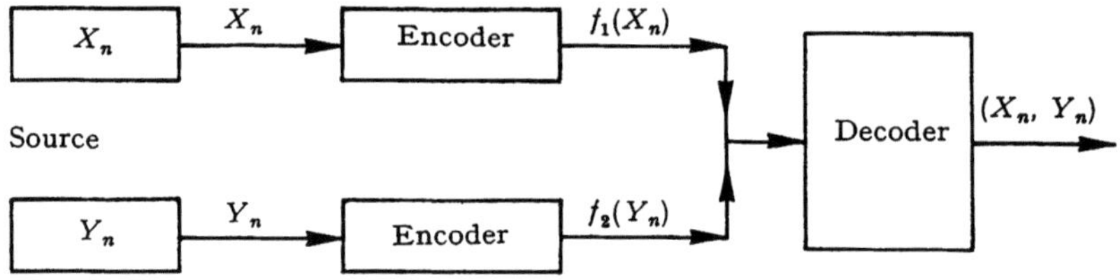

Our next theorem describes the achievable region, i.e., the set of achievable points.

Theorem 12.1.2. The achievable region for the problem of this section is described in the following diagram:

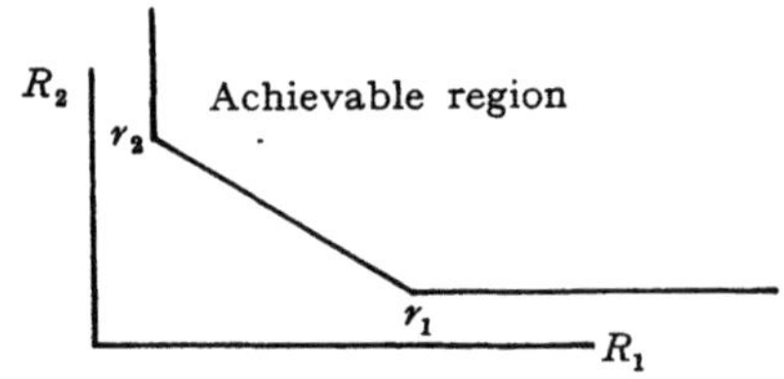

Proof: What remains to be proved is that the horizontal and vertical lines are in the boundary of the achievable region. The proof being essentially the same for both, it will be enough to prove the following: Even if $R_1 > H(X)$, R_2 cannot be $< H(Y|X)$.

In order to cover the space, the encoder must enumerate (encode) the π-sequences. Each of these generates at least $\exp_2\{nH(Y|X) - \sqrt{n}K(\lambda)\}$ sequences. Even if the sets of sequences generated by each π-sequence were disjoint, the rate for the encoder of the sequences in B_n^* would have to be, for large n, close to $H(Y|X)$. This is a fortiori true since the sets of generated sequences are not disjoint.

12.2. Source coding with side information at the decoder. The problem of this section will be described precisely in a moment, but it can be readily understood from the following diagram.

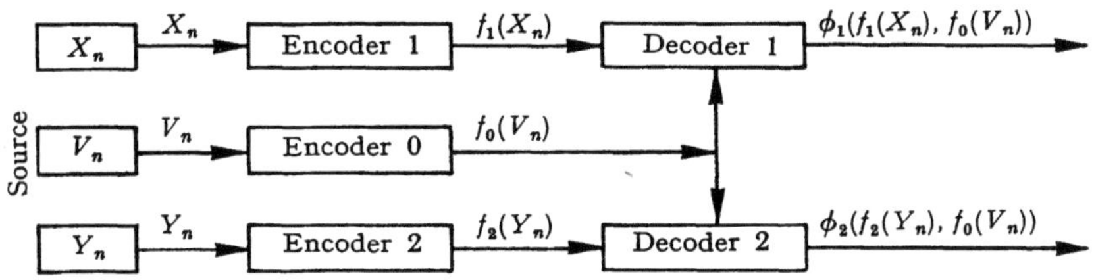

The chance variables (X^i, Y^i, V^i) have a (joint) distribution which is the same for all i, and are independently distributed for different i. V^i takes the values $1, \ldots, c$; (X^i, Y^i) is distributed as in the preceding section. X_n and Y_n are as defined in the preceding section, and $V_n = (V^1, \ldots, V^n)$. Encoders 1, 0, 2 receive, respectively, X_n, V_n, and Y_n, and produce the encoded numbers $f_1(X_n)$, $f_0(V_n)$, $f_2(Y_n)$, respectively. Decoder 1 (resp. 2) uses the decoding function ϕ_1 (resp., ϕ_2) so that, with suitably large probability,

$$\phi_1(f_1(X_n), f_0(V_n)) = X_n \tag{12.2.1}$$

(respectively,

$$\phi_2(f_2(Y_n), f_0(V_n)) = Y_n) . \tag{12.2.2}$$

The spaces A_n^* and B_n^* must, of course, be covered. The problem is to determine the region $\bar{R}$ of achievable rate triples (R_1, R_0, R_2). It is obvious that the region $\bar{R}$ is closed, convex, and M-closed.

Let X, Y, V be chance variables with the same distribution as X^1, Y^1, V^1. Let W be any chance variable with values in $\{1, \ldots, d\}$, such that (X, Y), V, W in this order form a Markov chain. We will prove

Theorem 12.2.1. For any such W and any n and $\lambda > 0$, there exist coding schemes for the problem of this section such that, defining

$$M_i = \|f_i\|, \; i = 0, 1, 2,$$

we have

$$M_0 \leqq \exp_2\{n\bar{I}(V, W) + \sqrt{n}K(\lambda)\}, \tag{12.2.3}$$

$$M_1 \leqq \exp_2\{nH(X|W) + \sqrt{n}K(\lambda)\} \tag{12.2.4}$$

and

$$M_2 \leqq \exp_2\{nH(Y \mid W) + \sqrt{n}K(\lambda)\}. \qquad (12.2.5)$$

Corollary. Let R^* be the M-closure of the union over all d and W of the triples

$$\bar{I}(V, W), H(X \mid W), H(Y \mid W). \qquad (12.2.6)$$

Then $\bar{R} \supseteq R^*$.

Proof of the theorem: Using Lemma 11.2.1 we can obtain at most

$$M_0 = \exp_2\{n[\bar{I}(V, W)] + \sqrt{n}K(\lambda)\} \qquad (12.2.7)$$

disjoint sets A_i of sequences v_n which cover the space of all sequences v_n. These are the decoding sets of a code whose transmitted words are μn-sequences in the alphabet $\{1, \ldots, d\}$, where μ is the distribution of W. Call these words $\alpha_1, \ldots \alpha_{M_0}$. The sets $A_1, \ldots, A_{M_0}$ consist only of n-sequences generated $(W \to V)$ by the α_1. The indices of the α_i are the values of our function f_0.

If the multiplicative constants are chosen properly then generation in a Markov chain is a transitive operation in the following sense: If a fixed n-sequence in the W alphabet generates $(W \to V)$ a fixed n-sequence in the V alphabet, and the latter generates $(V \to X)$ a chance n-sequence in the X alphabet, the first n-sequence generates $(W \to X)$ the last n-sequence, with a probability which can be made as close to one as desired. All the n-sequences in the X alphabet which are generated by n-sequences in A_i and are also generated $(W \to X)$ by α_i. are at most

$$\exp_2\{nH(X \mid W) + \sqrt{n}K(\lambda)\} \qquad (12.2.8)$$

in number. It follows from the proof of Theorem 12.1.1 that there exists an encoding function f_1 and a decoding function ϕ_1 with the desired properties. The existence of the desired f_2 and ϕ_2 is proved in the same way. This proves the theorem.

Let us prove that R^* is already obtained when $d = c + 2$. Let t be any distribution on the set $C = \{1, \ldots, c\}$; it is to be thought of as the set of transition probabilities from a fixed value of W, say w, to the c values of V. To each t there corresponds a vector $m(t)$ in $(c + 3)$ − space as follows: Its first a coordinates are the components of t. Its next three coordinates are

$$H(V \mid W = w), H(X \mid W = w), H(Y \mid W = w),$$

which are continuous functions of t. (The value of w obviously does not matter at all.) Let T be the set of points $m(t)$ for all t. Then T, as the continuous map of a compact, connected set, is itself compact and connected. It has dimension $(c + 2)$, since the first c coordinates always add to one. Let $\bar{T}$ be its convex hull. By the generalized Carathéodory theorem (EGGLESTON [1], Theorem 18), any point in $\bar{T}$ can be represented as a convex linear combination, with non-negative coefficients which sum to one, of $(c+2)$ points of T. The set $\bar{T}$ contains the point F, as follows: Its first c coordinates are the distribution of V; these determine $H(V)$. The next three coordinates are

$$H(V\,|\,W)\,,\ H(X\,|\,W)\,,\ H(Y\,|\,W)\ .$$

Thus the point F determines the point (12.2.6), and it suffices to take $d = c+2$.

This argument recurs several times in the text below. In order to avoid repeating it each time we will refer to it simply as the standard argument based on Carathéodory's theorem, and content ourselves with describing the essential mapping, which above was $m(t)$. The fact that one can bound d above has as a consequence that the region R^* can be computed to any desired degree of accuracy. We may not always mention this obvious consequence in the text which follows.

We now prove

Theorem 12.2.2. Let $\delta > 0$ be arbitrary. Let n be sufficiently large and $\lambda > 0$ be sufficiently small. Let f_0, f_1, f_2 be encoding functions for the problem of this section. Define $M_i = \|f_i\|$, $i = 0, 1, 2$. Then the rate triple

$$\frac{1}{n}\log M_0\,,\ \frac{1}{n}\log M_1\,,\ \frac{1}{n}\log M_2\,, \tag{12.2.9}$$

lies within a distance at most δ from R^*.

Proof: Define

$$S_0 = f_0(V_n)\,,\ S_1 = f_1(X_n)\,,\ S_2 = f_2(Y_n)\ . \tag{12.2.10}$$

Then

$$\log M_0 \geqq H(S_0) \geqq \bar{I}(S_0, V_n) \geqq H(V_n) - H(V_n\,|\,S_0)\ . \tag{12.2.11}$$

Also

$$H(V_n) = \sum_{i=1}^{n} H(V^i) \tag{12.2.12}$$

and

$$H(V_n\,|\,S_0) = \sum_{i=1}^{n} H(V^i\,|\,V_{i-1}, S_0)\ . \tag{12.2.13}$$

Write $W^i = (V_{i-1}, S_0)$. Hence

$$\log M_0 \geqq \sum_{i=1}^{n} \bar{I}(V^i, W^i)\ . \tag{12.2.14}$$

It is easy to verify that Theorem 7.4.1 holds for any distribution Q_0' on G_1, not only for the uniform distribution, provided only that one replaces $\bar{\lambda}$ in (7.4.1) by the expected value of the error with respect to the new Q_0'. To avoid the necessity for more notation call this new value also $\bar{\lambda}$. From this we obtain, for $1 \leqq j \leqq M_0$,

$$H(X_n\,|\,\phi_1(S_0, S_1)) \leqq \bar{\lambda} n \log a + 1\,, \tag{12.2.15}$$

and

$$H(Y_n\,|\,\phi_2(S_0, S_2)) \leqq \bar{\lambda} n \log b + 1\ . \tag{12.2.16}$$

When S_0 is given, $\phi_i(S_0, S_i)$ can take no more than M_i different values, $i = 1, 2$. Hence

$$\begin{aligned}\log M_1 &\geqq H(\phi_1(S_0, S_1)\,|\,S_0)\\ &\geqq \bar{I}(X_n, \phi_1(S_0, S_1)\,|\,S_0)\\ &= H(X_n|S_0) - H(X_n|S_0, \phi_1(S_0, S_1))\\ &\geqq H(X_n|S_0) - H(X_n|\phi_1(S_0, S_1))\\ &\geqq H(X_n|S_0) - 1 - \bar{\lambda} n \log a\,.\end{aligned} \tag{12.2.17}$$

Similarly

$$\log M_2 \geqq H(Y_n|S_0) - 1 - \bar{\lambda} n \log b\,. \tag{12.2.18}$$

Also

$$\begin{aligned}H(X_n|S_0) &= \sum_{i=1}^{n} H(X^i|X_{i-1}, S_0)\\ &\geqq \sum_{i=1}^{n} H(X^i|X_{i-1}, V_{i-1}, S_0)\\ &= \sum_{i=1}^{n} H(X^i|V_{i-1}, S_0)\end{aligned} \tag{12.2.19}$$

(because X_{i-1}, (V_{i-1}, S_0), X^i in this order form a Markov chain). Hence

$$H(X_n|S_0) = \sum_{i=1}^{n} H(X^i|W^i). \tag{12.2.20}$$

Similarly

$$H(Y_n|S_0) \geq \sum_{i=1}^{n} H(Y^i|W^i) \tag{12.2.21}$$

Substituting (12.2.20) into (12.2.17), we obtain

$$\log M_1 \geqq \sum_{i=1}^{n} H(X^i|W^i) - 1 - \bar{\lambda} n \log a\,. \tag{12.2.22}$$

Similarly,

$$\log M_2 \geqq \sum_{i=1}^{n} H(Y^i|W^i) - 1 - \bar{\lambda} n \log b\,. \tag{12.2.23}$$

We now note that

$$(X^i, Y^i),\ V^i,\ W^i$$

in this order form a Markov chain. Hence the rate triple

$$\bar{I}(V^i, W^i),\ H(X^i|W^i),\ H(Y^i|W^i) \tag{12.2.24}$$

belongs to R^* for every i, by the definition of R^*. From the fact that R^* is M-closed, (12.2.24), (12.2.14), (12.2.22), and (12.2.23), the rate triple

$$\frac{1}{n}\log M_0,\ \frac{1}{n}\log M_1 + \bar{\lambda}\log a + \frac{1}{n},\ \frac{1}{n}\log M_2 + \bar{\lambda}\log b + \frac{1}{n}$$

is in R^*. When $\bar{\lambda}$ is sufficiently small and n sufficiently large, the rate triple

$$\frac{1}{n}\log M_0,\ \frac{1}{n}\log M_1,\ \frac{1}{n}\log M_2$$

either lies in R^* or can be made to lie arbitrarily closely to R^*. This proves the theorem.

12.3. Encoding assisted by a common channel. The problem of this section will be described precisely in a moment, but it can be readily understood from the following diagram.

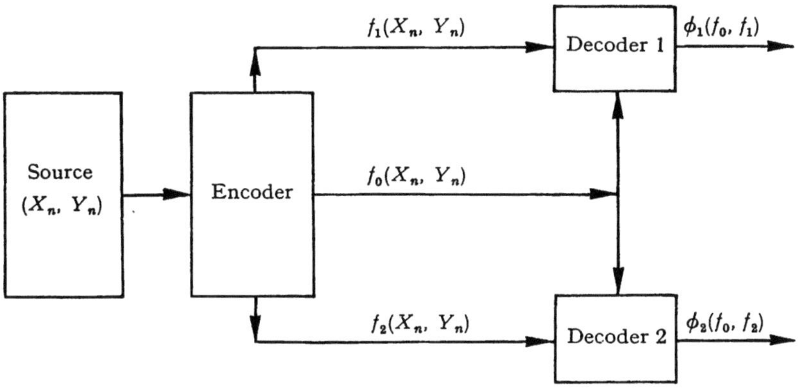

The chance variables X_n, Y_n are defined as in Section 12.1. A pair (X_n, Y_n) enters the encoder and is then encoded into three separate numbers $f_0(X_n, Y_n), f_1(X_n, Y_n), f_2(X_n, Y_n)$. Decoder 1 receives f_0 and f_1, decoder 2 receives f_0 and f_2. The decoding functions ϕ_i, $i = 1, 2$, (of the ith decoder) are to satisfy, with suitably large probability,

$$\phi_1(f_0(X_n, Y_n), f_1(X_n, Y_n)) = X_n \tag{12.3.1}$$

and

$$\phi_2(f_0(X_n, Y_n), f_2(X_n, Y_n) = Y_n . \tag{12.3.2}$$

The spaces A_n^* and B_n^* must of course be covered. Our problem is to determine the region $\bar{R}$ of achievable rate triples R_0, R_1, R_2. It is obvious that $\bar{R}$ is closed, convex, and M-closed.

Let (X, Y) be chance variables with the same distribution as (X^1, Y^1). Let W be any chance variable with values in $\{1, \ldots, d\}$, defined on the same probability space as (X, Y). (Consequently the joint distribution of X, Y, W has a marginal distribution which is the given distribution of (X, Y).) As usual, define $M_i = \|f_i\|$, $i = 0, 1, 2$.

Theorem 12.3.1. For any such W and any n and $\lambda > 0$, there exist coding schemes for the problem of this section such that

$$M_0 \leq \exp_2\{n\bar{I}(W, [X, Y]) + \sqrt{n}K(\lambda)\} , \tag{12.3.3}$$

$$M_1 \leq \exp_2\{n(H(X|W) + \sqrt{n}K(\lambda)\} , \tag{12.3.4}$$

$$M_2 \leq \exp_2\{nH(Y|W) + \sqrt{n}K(\lambda)\} . \tag{12.3.5}$$

The proof of this theorem differs only in minor details, but not in any essential idea, from the proof of Theorem 12.2.1. We therefore omit it. (See also Section 13.1.). Just as in the proof of Theorem 12.2.1, it can be shown that it suffices to take $d = ab + 2$.

Let R^* be the M-closure of the union over all W of the rate triples

$$\bar{I}(W, [X, Y]), H(X \mid W), H(Y \mid W). \tag{12.3.6}$$

We now prove

Theorem 12.3.2. Let $\delta > 0$ be arbitrary. Let n be sufficiently large and $\lambda > 0$ be sufficiently small. Let f_0, f_1, f_2 be the encoding functions for a problem of this section. Define $M_i = \|f_i\|$, $i = 0, 1, 2$. Then the rate triple

$$\frac{1}{n}\log M_0, \frac{1}{n}\log M_1, \frac{1}{n}\log M_2 \tag{12.3.7}$$

lies within a distance at most δ from R^*.

Proof: Define $S_i = f_i(X_n, Y_n)$, $i = 0, 1, 2$, and $W^i = (S_0, X_{i-1}, Y_{i-1})$, $i = 1, \ldots, n$. Then, exactly as in the proof of Theorem 12.2.2., we obtain

$$\log M_0 \geqq H(S_0) \geqq \bar{I}(S_0, [X_n, Y_n]) \tag{12.3.8}$$

$$\geqq \sum_{i=1}^{n} \bar{I}(W^i, [X^i, Y^i])$$

$$\log M_1 \geqq H(X_n \mid S_0) - 1 - \bar{\lambda} n \log a \tag{12.3.9}$$

$$\log M_2 \geqq H(Y_n \mid S_0) - 1 - \bar{\lambda} n \log b. \tag{12.3.10}$$

Also

$$H(X_n \mid S_0) = \sum_{i=1}^{n} H(X^i \mid X_{i-1}, S_0) \geqq \sum_{i=1}^{n} H(X^i \mid W^i) \tag{12.3.11}$$

$$H(Y_n \mid S_0) \geq \sum_{i=1}^{n} H(Y^i \mid W_i) \tag{12.3.12}$$

Hence

$$\log M_1 \geqq \sum_{i=1}^{n} H(X^i \mid W^i) - 1 - \bar{\lambda} n \log a \tag{12.3.13}$$

$$\log M_2 \geqq \sum_{i=1}^{n} H(Y^i \mid W^i) - 1 - \bar{\lambda} n \log b. \tag{12.3.14}$$

The proof is now completed as in Theorem 12.2.2.

Remarks

Theorem 12.1.1, which initiated the subject of source coding, is due to Slepian and Wolf [1]. The proof given here is that of Cover [1]. The methods of Section 2.1 and Chapter 3 enable the improvement to $O(\sqrt{n})$ in the exponent.

Theorems 12.2.1 and 12.2.2, are due to Wyner [1], and Theorems 12.3.1 and 12.3.2 are due to Gray and Wyner [1]. The particularly simple proofs of Theorems 12.2.1 and 12.3.1 given here are new, as is the improvement to terms $O(\sqrt{n})$ in the exponent. The proofs of Theorems 12.2.2 and 12.3.2 are also particularly simple and are due to the respective authors.

Ahlswede and Körner [1] proved Theorems 12.2.1 and 12.2.2 for the problem of Section 12.2, simplified by the delection of the Y's and decoder 2. Their proofs are more complicated. For this problem Ahlswede, Gacs, and Körner [1], [2], proved the strong version of Theorem 12.2.2, which, together with Theorem 12.2.1, asserts that $\bar{R} = R^*$. This more difficult result uses their Theorem 7.9.2.

13. Source Coding and Rate Distortion

13.1. The problem of Section 12.3 for rate distortion. We now consider the problem of Section 12.3, whose notation we also adopt, except that now

$$\hat{X}_n = \phi_1(f_0(X_n, Y_n), f_1(X_n, Y_n))$$

is an n-sequence in the reproduction alphabet $\{1,\ldots,c\}$. As usual, C_n^* will denote the space of such sequences. A distortion function d_1 between elements of $A = \{1,\ldots,a\}$ and $C = \{1,\ldots,c\}$ is given, and the distortion between n-sequences defined in the usual manner, i.e.,

$$d_1(a_n, c_n) = d_1([a^1,\ldots,a^n], [c^1,\ldots,c^n])$$

$$= \frac{1}{n}\sum_{i=1}^{n} d_1(a^i, c^i)\,.$$

Similarly,

$$\hat{Y}_n = \phi_2(f_0(X_n, Y_n), f_2(X_n, Y_n))$$

is an n-sequence in the reproduction alphabet $C' = \{1,\ldots,c'\}$. The space of such sequences will be denoted by C'^*_n. A distortion function d_2 between elements of $B = \{1,\ldots,b\}$ and those of C' is given, and d_2 over n-sequences defined in the usual way.

Let z_1 and z_2 be any numbers in suitable intervals as discussed in Section 11.1. Let $\varepsilon > 0$ be arbitrary. Suppose that the functions $f_0, f_1, f_2, \phi_1, \phi_2$ are such that

$$\mathrm{E}d_1(X_n, \hat{X}_n) < z_1 + \varepsilon\,, \tag{13.1.1}$$

and

$$\mathrm{E}d_2(Y_n, \hat{Y}_n) < z_2 + \varepsilon\,. \tag{13.1.2}$$

Of course, f_0, f_1, f_2, ϕ_1 and ϕ_2 depend upon n and ε. Define the rate triple

$$\frac{1}{n}\log M_i = \frac{1}{n}\log\|f_i\|\,,\ i = 0, 1, 2\,. \tag{13.1.3}$$

A rate triple (r_0, r_1, r_2) is called achievable with respect to (z_1, z_2), or simply achievable if there is no doubt about which (z_1, z_2) is meant, if, for any ε and δ both positive and n sufficiently large, there exist functions $f_0, f_1, f_2, \phi_1, \phi_2$ such that (13.1.1) and (13.1.2) are satisfied, and

$$\frac{1}{n}\log\|f_1\| < r_i + \delta\,,\quad i = 0, 1, 2\,. \tag{13.1.4}$$

Our problem is to determine $\overline{R}(z_1, z_2)$, the set of all achievable rate triples, for any pair (z_1, z_2) in the appropriate interval. Obviously $\overline{R}(z_1, z_2)$ is closed, convex, and M-closed.

Let (X, Y) be chance variables with the same distribution as (X^1, Y^1). Let W be any chance variable with values in the set $\{1, \ldots, d\}$, defined on the same probability space as (X, Y). Let $\hat{X}$ be a chance variable, with values in C, defined on the same probability space as X, Y, W, and satisfying

$$\mathrm{E}d_1(X, \hat{X}) \leq z_1. \tag{13.1.5}$$

Recalling (11.5.14) we define

$$D_1(z_1 | W) = \min \bar{I}(X, \hat{X} | W), \tag{13.1.6}$$

where the minimum is with respect to all chance variables $\hat{X}$ which satisfy the above requirements. Also define

$$D_2(z_2 | W) = \min \bar{I}(Y, \hat{Y} | W), \tag{13.1.7}$$

where $\hat{Y}$ is a chance variable, with values in C', defined on the same probability space as $X, Y, W, \hat{X}$, and satisfying

$$\mathrm{E}d_2(Y, \hat{Y}) \leqq z_2. \tag{13.1.8}$$

The minimum is with respect to all chance variables $\hat{Y}$ which satisfy the above requirements.

Let $R^*(z_1, z_2)$ be the M-closure of the set of triples (r_0, r_1, r_2) such that

$$r_0 = \bar{I}(W, [X, Y]), \tag{13.1.9}$$

$$r_1 = D_1(z_1 | W), \tag{13.1.10}$$

$$r_2 = D_2(z_2 | W), \tag{13.1.11}$$

for all d and W which satisfy the above requirements. We now prove

Theorem 13.1.1. We have, for all (z_1, z_2) in the proper interval,

$$\bar{R}(z_1, z_2) \supseteq R^*(z_1, z_2). \tag{13.1.12}$$

The proof of this result is so simple, like that of Theorem 12.3.1, that it will be sufficient to sketch it without filling in all the obvious details. Let n be sufficiently large. As in the proof of Theorem 12.3.1, construct a code (n, N, ε') whose message sequences $\alpha_1, \ldots, \alpha_N$ are $\pi_0 n$-sequences, where π_0 is the distribution of W, whose decoding sets $G_1, \ldots, G_N$ are disjoint sets of sequences in $A_n^* \times B_n^*$, each G_i consists only of sequences generated $(W \to (X, Y))$ by α_i, $\log N$ is approximately (we shall omit the terms of lower order in the exponent)

$$n\bar{I}(W, [X, Y]), \quad \text{and} \quad P\{(X_n, Y_n) \,\varepsilon \bigcup_i G_i\} > 1 - \varepsilon''.$$

Consider one of the message sequences, say α_i. The frequency of the element j in α_i is approximately $n\pi_{0j}$. Consequently, the frequency of the various values of X, in any sequence in G_i, in those places which correspond to values j of α_i, is approximately that given by the conditional distribution of X, given $W = j$. It follows from the definition of $D_1(\cdot \mid W)$ ((13.1.6)) that, *on* G_i, one can define a function f_1, say, such that, for n sufficiently large, f_1 (which is really a function only of the x_n-sequence in a sequence (x_n, y_n) in G_i) will take at most

$$\exp_2\{n D_1(z_1 \mid W)\}$$

values in C_n^* which span the x_n-sequences in G_i to within distortion $z_1 + \varepsilon$.

The above construction defines f_1 only on G_i, and the decoder has to know the value i. This will be furnished by the value of $f_0(x_n, y_n)$, which we define as follows: If $(x_n, y_n) \varepsilon G_i$, $f_0(x_n, y_n) = i$. If $(x_n, y_n) \notin \bigcup G_i$, define $f_0(x_n, y_n) = 1$ arbitrarily. (The actual value then does not matter.) The procedure for f_2 is similar. This proves the theorem.

Theorem 13.1.2. The following inequality holds for all (z_1, z_2):

$$R(z_1, z_2) \leqq R^*(z_1, z_2) . \tag{13.1.13}$$

Proof: For typographical simplicity write

$$S_0 = f_0(X_n, Y_n) , \quad S_1 = f_1(X_n, Y_n) , \quad S_2 = f_2(X_n, Y_n) ,$$
$$M_0 = \|f_0\| , \quad M_1 = \|f_1\| , \quad M_2 = \|f_2\| .$$

Also write

$$\phi_1(f_0(X_n, Y_n), f_1(X_n, Y_n)) = \hat{X}_n = (\hat{X}^1, \ldots, \hat{X}^n)$$
$$\phi_2(f_0(X_n, Y_n), f_2(X_n, Y_n)) = \hat{Y}_n = (\hat{Y}^1, \ldots, \hat{Y}^n)$$
$$\mathrm{E} d_1(X_n, \hat{X}_n) = \frac{1}{n} \sum_{i=1}^{n} \mathrm{E} d_1(X^i, \hat{X}^i) = z_1$$
$$\mathrm{E} d_2(Y_n, \hat{Y}_n) = \frac{1}{n} \sum_{i=1}^{n} \mathrm{E} d_2(Y^i, \hat{Y}^i) = z_2$$
$$V^i = (S_0, X_{i-1}) , \; W^i = (S_0, X_{i-1}, Y_{i-1}) .$$

Let X, Y, V, W be chance variables defined on the same sample space as (X_n, Y_n), with the distribution given by

$$P\{(X, Y, V, W) = (X^i, Y^i, V^i, W^i)\} = \frac{1}{n}, \; i = 1, \ldots, n . \tag{13.1.14}$$

Thus (X, Y) has the same distribution as (X^1, Y^1). Let $\hat{X}$ be a chance variable, with values in C, defined on the same sample space as above, and such that, for any z in the proper interval,

$$\mathrm{E} d_1(X, \hat{X}) \leqq z . \tag{13.1.15}$$

Then define

$$D_1(z \mid V) = \min \bar{I}(X, \hat{X} \mid V) , \tag{13.1.16}$$

$$D_1(z \mid W) = \min \bar{I}(X, \hat{X} \mid W) , \tag{13.1.17}$$

where each minimum is with respect to all chance variables $\hat{X}$ which satisfy (13.1.15). Since knowledge of W includes knowledge of V we have, from (11.5.14), that, for every z,

$$D_1(z \mid W) \leqq D_1(z \mid V) . \tag{13.1.18}$$

We have, from the definition of M_0,

$$\begin{aligned}
\frac{1}{n} \log M_0 &\geqq \frac{1}{n} H(S_0) \geqq \frac{1}{n} \bar{I}(S_0, [X_n, Y_n]) \\
&= \frac{1}{n}[H(X_n, Y_n) - H(X_n, Y_n \mid S_0)] \qquad (13.1.19)\\
&= \frac{1}{n} \sum_{i=1}^{n} [H(X^i, Y^i) - H(X^i, Y^i \mid W^i)]
\end{aligned}$$

(this step by (2.2.2))

$$= \bar{I}([X, Y], W) .$$

When S_0 is fixed, ϕ_1 cannot take more than M_1 values. Hence

$$\begin{aligned}
\frac{1}{n} \log M_1 &\geqq \frac{1}{n} H(\hat{X}_n \mid S_0) \\
&\geqq \frac{1}{n} \sum_{i=1}^{n} [H(\hat{X}_n \mid S_0, X_{i-1}) - H(\hat{X}_n \mid S_0, X_i)] \\
&= \frac{1}{n} \sum_{i=1}^{n} \bar{I}(\hat{X}_n, X^i \mid V^i) \\
&= \frac{1}{n} \sum_{i=1}^{n} [H(X^i \mid V^i) - H(X^i \mid V^i, \hat{X}_n)] \qquad (13.1.20)\\
&\geq \frac{1}{n} \sum_{i=1}^{n} [H(X^i \mid V^i) - H(X^i \mid V^i, \hat{X}^i)] \\
&= \frac{1}{n} \sum_{i=1}^{n} \bar{I}(X^i, \hat{X}^i \mid V^i) \\
&\geqq D_1(z_1 \mid V) \quad \text{(by definition of } D_1(\cdot \mid V)) \\
&\geqq D_1(z_1 \mid W) \quad \text{(by (13.1.18))} .
\end{aligned}$$

In essentially the same way a similar inequality is proved for $\frac{1}{n}\log M_2$. From the latter, (13.1.19), and (13.1.20), the theorem follows. Theorems 13.1.1 and 13.1.2 are due to GRAY and WYNER [1].

For any value w of W consider the following vector m: Its first $abcc'$ coordinates are

$$P\{X = s_1,\ Y = s_2,\ \hat{X} = s_3,\ \hat{Y} = s_4 \mid W = w\} \tag{13.1.21}$$

$$1 \leq s_1 \leq a,\ 1 \leq s_2 \leq b,\ 1 \leq s_3 \leq c,\ 1 \leq s_4 \leq c' .$$

For each w these coordinates add to one. Its next coordinate is $H(X, Y \mid W = w)$. The next two are $H(X \mid W = w)$, $H(X \mid \hat{X}, W = w)$. The last two are $H(Y \mid W = w)$, $H(Y \mid \hat{Y}, W = w)$. Thus the last five coordinates are functions of the first $abcc'$. By the standard argument based on Carathéodory's theorem we conclude that R^* is already obtained when $d = abcc' + 4$, and can therefore be computed to any degree of accuracy.

13.2. The rate distortion function for source coding with side information at the decoder. This section is devoted to the proof of Theorem 11.5.1. We adopt the notation of Section 11.5. For convenience we divide the proof into two parts. Until the end we assume that Z takes values in the set $S = \{1, \ldots, s\}$. Then we will show that it suffices to take $s = a + 1$.

Theorem 13.2.1. Let $\varepsilon > 0$ be arbitrary but fixed. There exists a sequence $\{f_n\}$ such that

$$\overline{\lim_{n\to\infty}}\ \frac{1}{n}\log \|f_n\| \leqq \min\,[\bar{I}(X, Z) - \bar{I}(Y, Z)] . \tag{13.2.1}$$

The minimum is over all chance variables Z which satisfy the conditions of Theorem 11.5.1, except that $S = \{1, \ldots, s\}$.

Proof: Fix some Z which satisfies the required conditions, and let π be its distribution. Let (Y^i, X^i, Z^i), $i = 1, 2, \ldots$ be independent chance variables with the same distribution as (Y, X, Z). Write $Z_n = (Z^1, \ldots, Z^n)$. Let n_0 be sufficiently large. Define

$$\psi_n(a_n, z_n) = (\psi(a^1, z^1), \ldots, \psi(a^n, z^n)) .$$

We again use Lemma 11.2.1, the device by which we proved Theorem 11.2.1, and the transitivity of generation discussed in the paragraph which contains (12.2.8). Let $(\alpha_1, G_1), \ldots, (\alpha_N, G_N)$ be a code (n_0, N, λ) such that the α's are πn_0-sequences in the alphabet S, each G_i consists only of sequences in $A^*_{n_0}$ which are generated by α_i, and

$$P\left\{X_{n_0} \varepsilon \bigcup_{i=1}^{N} G_i\right\} > 1 - \varepsilon'', \text{ say.} \tag{13.2.2}$$

Then

$$N < \exp_2\{n_0 \bar{I}(X, Z) + \sqrt{n_0} K_1\}, \tag{13.2.3}$$

where K_1 is, and every other K will be, a suitable positive constant.

Let L_i be the set of sequences in B^*_n which are generated by the sequences in G_i. If the multiplicative constants are properly chosen, with probability arbitrarily close to one, a sequence in L_i is generated by α_i. From this we draw the following conclusion: Let α_{ik} be the kth letter of α_i. Define the chance variable μ_k as the kth letter of the chance sequence V generated $Z \to X$ by α_i, and the chance variable μ'_k as the kth letter of the chance sequence V' generated $X \to Y$ by V. Then the average, over k, distribution of $(\alpha_{ik}, \mu_k, \mu'_k)$ can be made arbitrarily close to the distribution of (Z, X, Y) by making n_0 sufficiently large.

Define the function g as follows: For every sequence a_n in G_i, $g(a_n) = \alpha_i$. If $a_n \notin \bigcup_i G_i$, $g(a_n) = a_1$, say. From the previous paragraph, the definition of ψ, and the law of large numbers, we obtain, when n_0 is sufficiently large, that

$$\mathrm{E} d\left(X_{n_0}, \psi_{n_0}\left(Y_{n_0}, g(X_{n_0})\right)\right) < z + \varepsilon'. \tag{13.2.4}$$

This has the following consequence: If the encoder were to send out, as $f_{n_0}(X_{n_0})$, the *index* of the α sequence which is the value of $g(X_{n_0})$, the decoder would know $g(X_{n_0})$, and be able to construct $\psi_{n_0}(Y_{n_0}, g(X_{n_0}))$, which would be his γ_{n_0} and fulfill the requirements (11.5.3). The trouble is that the N of (13.2.3) is much too large and does not satisfy (13.2.1). We now proceed to reduce N by the use of Theorem 12.1.1.

Let $n = mn_0$, where the integer m is sufficiently large. Let $X(i)$, $i = 1, \ldots, m$, be successive blocks of X's, each of length n_0. Let β_i be the index of the α sequence which is the value of $g(X(i))$, and let β be a chance variable with the same distribution as each of the β_i. It follows from Theorem 12.1.1 that, if

$$\log \|f_n\| = mH(\beta \mid Y_{n_0}) + \sqrt{m} K_2, \tag{13.2.5}$$

the decoder, who knows Y_n, will be able to reconstruct, with suitably high probability, the sequence

$$\beta_1, \ldots\ldots, \beta_m \tag{13.2.6}$$

and hence the sequence

$$g(X(1)),\ldots,g(X(m))\,. \tag{13.2.7}$$

The desired function

$$\gamma_n(f_n(X_n),\,Y_n)=\psi_n(Y_n,\,[g(X(1)),\ldots,g(X(m))] \tag{13.2.8}$$

satisfies (13.2.4) with n_0 replaced by n. It remains to evaluate the right member of (13.2.5).

We have

$$\begin{aligned}H(\beta\,|\,Y_{n_0})&=H(\beta,\,Y_{n_0})-H(Y_{n_0})\\&=H(\beta)-n_0H(Y)+H(Y_{n_0}\,|\,\beta)\,.\end{aligned} \tag{13.2.9}$$

From (13.2.3),

$$H(\beta)<n_0\bar{I}(X,Z)+\sqrt{n_0}K_1. \tag{13.2.10}$$

From Lemma 2.1.5 and the construction of the code $(\alpha_1,G_1),\ldots,(\alpha_N,G_N)$ we obtain that

$$H(Y_{n_0}\,|\,\beta)<n_0H(Y\,|\,Z)+\sqrt{n_0}K_3. \tag{13.2.11}$$

From (13.2.5) and (13.2.9)–(13.2.11) we obtain that

$$\log\|f_n\|<n\,[\bar{I}(X,Z)-\bar{I}(Y,Z)]+m\sqrt{n_0}K_4. \tag{13.2.12}$$

By making n_0 large enough we obtain (13.2.1), provided $\varepsilon'<\varepsilon$.

It remains only to consider the case when $n=mn_0+m'$, $0<m'<n_0$, i.e., when n is not an integral multiple of n_0. In that case we simply "waste" the last (say) m' observations. If $\varepsilon'<\varepsilon$, (11.5.3) will hold. This completes the proof of Theorem 13.2.1.

Theorem 13.2.2. In the problem of Theorem 11.5.1 we have, for every n,

$$\frac{1}{n}\log\|f_n\|\geq\min\,[\bar{I}(X,Z)-\bar{I}(Y,Z)]\,. \tag{13.2.13}$$

Here the minimum is over chance variables Z such that Y, X, Z form a Markov chain and there exists a function ψ such that

$$\mathrm{E}d(X,\,\psi(Y,Z))\leq z+\varepsilon\,.$$

(Note change from (11.5.4.) Z takes values in $S=\{1,\ldots,s\}$.

Proof: Write

$$\hat{X}_n=(\hat{X}^1,\ldots,\hat{X}^n)=\gamma_n(f_n(X_n),\,Y_n)\,,\qquad z_k=\mathrm{E}d(X^k,\hat{X}^k)\,,$$

so that

$$\sum_{k=1}^{n} z_k < n(z + \varepsilon) . \tag{13.2.14}$$

Now

$$\begin{aligned}
\log \|f_n\| &\geq H(f_n(X_n)) \geqq \bar{I}(f_n(X_n), X_n) \\
&= H(X_n) - H(X_n | f_n(X_n)) \\
&= H(X_n) - H(X_n | Y_n, f_n(X_n)) \\
&\quad - \bar{I}(X_n, Y_n | f_n(X_n)) \\
&= H(Y_n | X_n, f(X_n)) - H(Y_n | f_n(X_n)) \\
&\quad + H(X_n) - H(X_n | Y_n, f_n(X_n)) \\
&\geq H(Y_n | X_n) - H(Y_n) \\
&\quad + H(X_n) - H(X_n | Y_n, f_n(X_n)) .
\end{aligned} \tag{13.2.15}$$

Hence

$$\begin{aligned}
\log \|f_n\| \geq &- \bar{I}(X_n, Y_n) \\
&+ \sum_{i=1}^{n} [H(X^i) - H(X^i | Y_n, f_n(X_n), X_{i-1})] .
\end{aligned} \tag{13.2.16}$$

Define

$$Z^i = (X_{i-1}, Y^1, \ldots, Y^{i-1}, Y^{i+1}, \ldots, Y^n, f_n(X_n)) . \tag{13.2.17}$$

Then, from (13.2.16),

$$\begin{aligned}
\log \|f_n\| \geqq \sum_{i=1}^{n} [&- \bar{I}(X^i, Y^i) \\
&+ H(X^i) - H(X^i | Z^i, Y^i)] .
\end{aligned} \tag{13.2.18}$$

From the definition of Z^i it follows that Z^i, X^i, Y^i, in this order, form a Markov chain. $\hat{X}_n$, and hence $\hat{X}^i$, are functions of (Y^i, Z^i). Write

$$\hat{X}^i = \psi_i(Y^i, Z^i) . \tag{13.2.19}$$

Let $J(z)$ be the right member of (11.5.5) (with condition (11.5.4)). Then

$$\bar{I}(X^i, Z^i) - \bar{I}(Y^i, Z^i) \geqq J(z_k) . \tag{13.2.20}$$

From the fact that Z^i, X^i, Y^i is a Markov chain we have that the right member of (13.2.18) equals

$$\begin{aligned}
&\sum_{i=1}^{n} [H(X^i|Y^i) - H(X^i, Y^i|Z^i) + H(Y^i|Z^i)] \\
&= \sum_{i=1}^{n} [H(X^i|Y^i) - H(X^i|Z^i) - H(Y^i|X^i, Z^i) \\
&\quad + H(Y^i|Z^i)] \\
&= \sum_{i=1}^{n} [H(X^i|Y^i) - H(X^i|Z^i) - H(Y^i|X^i) \\
&\quad + H(Y^i|Z^i)] \qquad (13.2.21) \\
&= \sum_{i=1}^{n} [\{H(X^i) - H(Y^i) + H(Y^i|X^i)\} - H(X^i|Z^i) \\
&\quad - H(Y^i|X^i) + H(Y^i|Z^i)] \\
&= \sum_{i=1}^{n} [H(X^i) - H(Y^i) - H(X^i|Z^i) + H(Y^i|Z^i)] \\
&= \sum_{i=1}^{n} [\bar{I}(X^i, Z^i) - \bar{I}(Y^i, Z^i)] \\
&\geqq \sum_{i=1}^{n} J(z_i) .
\end{aligned}$$

Temporarily assume the truth of the following

Lemma 13.2.1. The function J is convex.

From this lemma it follows at once that the right member of (13.2.21) is not less than

$$nJ\left(\frac{1}{n}\sum_{i=1}^{n} z_i\right) > nJ(z+\varepsilon) . \qquad (13.2.22)$$

This inequality follows from (13.2.14) and the fact that J is decreasing. This proves Theorem 13.2.2.

Theorem 11.5.1 follows from (11.5.2), Theorems 13.2.1 and 13.2.2, and the fact that J is decreasing and continuous (from Lemma 13.2.1). It remains only to prove Lemma 13.2.1.

Proof of Lemma 13.2.1. Write

$$\begin{aligned}
\mu(i) &= P\{Z=i\}, \quad t(j|i) = P\{X=j|Z=i\} \\
\pi(i) &= P\{X=i\}, \quad w(j|i) = P\{Y=j|X=i\} .
\end{aligned} \qquad (13.2.23)$$

The values of all these quantities determine the distribution of (Z, X, Y).

Also write

$$q(k\mid i) = P\{Y = k\mid Z = i\}$$

which we know to be equal to

$$\sum_j \mu(i)t(j\mid i)w(k\mid j)\,. \tag{13.2.24}$$

We have

$$\begin{aligned} &\mathrm{E}d(X, \psi(Y, Z)) \\ &= \sum_i \mu(i)\Big[\sum_{j,k} t(j\mid i)\; w(k\mid j)\; d(j,\psi(k,i))\Big] \\ &\geqq \sum_{i,k} \mu(i)\Big[\sum_j t(j\mid i)w(k\mid j)d(j,\;\psi_0(k,\,i))\Big], \end{aligned} \tag{13.2.25}$$

where ψ_0 is a function such that

$$d(j,\,\psi_0(k,\,i)) = \min_g d(j,g)\,. \tag{13.2.26}$$

The function ψ_0 is a function of the c.p.f. t but not of the distribution μ. Now

$$\bar{I}(X,Z) \dot{-} \bar{I}(Y,Z) = H(X) - H(Y) + H(Y\mid Z) - H(X\mid Z)\,.$$

The first two terms of the right member are constants. We may therefore restate the problem as follows: Define

$$J^*(z) = \min \sum_i \mu(i)\;[H(Y\mid Z = i) - H(X\mid Z = i)]\,. \tag{13.2.27}$$

The minimum is with respect to μ and t, subject to the condition that the right member of (13.2.25) equals z. We have to prove that J^* is convex. [Note that, for each i, the quantity in square brackets in (13.2.27) is a convex function of t (argument above Lemma 6.4.2), and t is a function of z. However μ is also a function of z, so this argument alone will not suffice.]

Consider the totality T of all vectors m whose first a coordinates are a c.p.f. $t(j\mid\alpha), j = 1,\ldots, a$, and whose next two (and last) coordinates are

$$\Big[\sum_{j,k} t(j\mid\alpha)w(k\mid j)d(j,\,\psi_0(k,\,\alpha))\Big]$$

and

$$[H\,(Y\mid Z = \alpha) - H\,(X\mid Z = \alpha)]\,.$$

T is the map of the compact, connected set of all vectors t under a continuous transformation. Hence T is compact and connected. Let $\bar{T}$ be its convex hull. The dimension of T is $a + 1$, since the first a coordinates always add to one. Since $\pi(j) = \sum_i \mu(i)t(j\mid i)$, there are points in $\bar{T}$

whose first a coordinates are the distribution π of X. The section S of $\bar{T}$, of all points with these first a coordinates, is also convex. Let S' be the convex subset of points of S whose next to last coordinate is z. Then $J^*(z)$ is the smallest last coordinate of the points of S'. Hence $J^*(\cdot)$ is a convex function of z.

By the generalized Carathéodory theorem (EGGLESTON [7], Theorem 18), every point of $\bar{T}$ can be represented as a convex combination of $a + 1$ points of T. This is therefore true for the point $\pi, z, J^*(z)$. This proves that it suffices to take $s = a + 1$.

14. Multiple Access Channels

14.1. Description of the problem. Consider the following diagram:

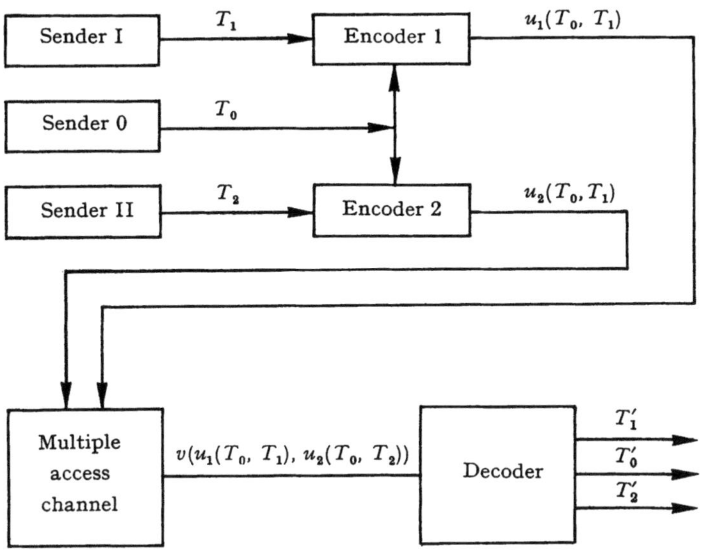

Let $A = \{1, \ldots, a\}$, $B = \{1, \ldots, b\}$, $C = \{1, \ldots, c\}$, A_n^* be the Cartesian product of n A's, and similarly for B_n^* and C_n^*. The multiple access channel is a kind of d.m.c. with c.p.f. $w(z \mid x, y)$, $x \varepsilon A$, $y \varepsilon B$, $z \varepsilon C$. Senders 0, I, II each independently choose at random one of M_0, M_1, M_2, respectively, equiprobable messages, say T_0, T_1, T_2. Encoder 1 then gives the function $u_1(T_0, T_1) \varepsilon A_n^*$, and encoder 2 gives the function $u_2(T_0, T_2) \varepsilon B_n^*$. These two n-sequences are then sent over the multiple access d.m.c. with c.p.f. w in the obvious way, and the chance sequence $v(u_1(T_0, T_1), u_2(T_0, T_2)) \varepsilon C_n^*$ is received by the decoder, which decodes it into a triple (T_0', T_1', T_2').

An error occurs if $T_i \neq T_i'$ for $i = 0$ or 1 or 2. Let $\bar{\lambda}$ denote the probability of an error. This is always a function of the code, i.e., the encoding and decoding procedures. We emphasize that T_0, T_1, and T_2 are chance variables, so it is really average error that is involved.

A rate triple (R_0, R_1, R_2) is called achievable or attainable if, for any ε and $\lambda > 0$, there exists an n_0 such that $n \geqq n_0$ implies that there exists a code with

$$M_i \geqq \exp_2\{n(R_i - \varepsilon)\},\ i = 0, 1, 2 \tag{14.1.1}$$

and

$$\bar{\lambda} \leqq \lambda. \tag{14.1.2}$$

We shall determine a region $\bar{R}$ such that any point in $\bar{R}$ is achievable, and such that, given any $\delta > 0$, the rate point

$$\left(\frac{1}{n}\log M_0,\ \frac{1}{n}\log M_1,\ \frac{1}{n}\log M_2\right)$$

for any code with small enough $\bar{\lambda}$ lies within a distance δ of $\bar{R}$.

Let $D = \{1, \ldots, d\}$, and U be a chance variable with distribution π on D. Let w_1 be a c.p.f. from D to A, and w_2 be a c.p.f. from D to B. Given U, X_1 and X_2 are to be independent chance variables with respective distributions w_1 and w_2. Given U, X_1 and X_2, Z is to be a chance variable with distribution $w(\cdot \mid X_1, X_2)$ Thus U, (X_1, X_2), Z form, in this order, a Markov chain.

Let $R^*(d, \pi, w_1, w_2)$ be the totality of all rate triples (R_0, R_1, R_2) which satisfy

$$0 \leqq R_1 \leqq \bar{I}(X_1, Z \mid X_2, U), \tag{14.1.3}$$

$$0 \leqq R_2 \leqq \bar{I}(X_2, Z \mid X_1, U), \tag{14.1.4}$$

$$0 \leqq R_1 + R_2 \leqq \bar{I}([X_1, X_2], Z \mid U), \tag{14.1.5}$$

$$0 \leqq R_0 + R_1 + R_2 \leqq \bar{I}([X_1, X_2], Z). \tag{14.1.6}$$

Let R^* be the m-closure of the union, over all d, π, w_1, w_2, of the $R^*(d, \pi, w_1, w_2)$.

14.2. A coding theorem. We now prove

Theorem 14.2.1. $\bar{R} \supseteq R^*$.

Proof: Fix d, π, w_1, and w_2. We use the method of random codes of Section 7.3 and Theorem 7.3.1. Let (U^i, X_1^i, X_2^i, Z^i), $i = 1, \ldots, n$, be independent chance variables with the same distribution as (U, X_1, X_2, Z). Write $U_n = (U^1, \ldots, U^n)$, $X_{1n} = (X_1^1, \ldots, X_1^n)$, and similarly for X_{2n} and Z_n. Let $U_n(i)$, $i = 1, \ldots, M_0$, be independent chance variables with the same distribution as U_n. When $U_n(i)$ is fixed let $X_{1n}(i, j_1)$, $X_{2n}(i, j_2)$, $j_1 = 1, \ldots, M_1$; $j_2 = 1, \ldots, M_2$, be independent chance variables. $X_{1n}(i, j_1)$

is to have the same distribution for all j_1; this distribution, conditional upon $U_n(i)$, is to be the same as the conditional distribution of X_{1n}, given U_n. Also $X_{2n}(i, j_2)$ is to have the same distribution for all j_2; this distribution, conditional upon $U_n(i)$, is to be the same as the conditional distribution of X_{2n}, given U_n.

A random code is constructed as follows: The sequence $u_1(i, j_1)$ is $X_{1n}(i, j_1)$, and the sequence $u_2(i, j_2)$ is $X_{2n}(i, j_2)$. Decoding is by the "strict maximum likelihood" method (Chapter 6 or (7.3.1)). We now compute $E\bar{\lambda}$, the expected value taken over the random codes. By symmetry, this is equal to

$$\begin{aligned} & EP\{T_0' \neq 1 \text{ or } T_1' \neq 1 \text{ or } T_2' \neq 1 \mid X_{1n}(1,1), X_{2n}(1,1)\} \\ &= EP\{T_0' = 1, T_1' \neq 1, T_2' = 1 \mid X_{1n}(1,1)\, X_{2n}(1,1)\} \\ &\quad + EP\{T_0' = 1, T_1' = 1, T_2' \neq 1 \mid X_{1n}(1,1), X_{2n}(1,1)\} \\ &\quad + EP\{T_0' = 1, T_1' \neq 1, T_2' \neq 1 \mid X_{1n}(1,1), X_{2n}(1,1)\} \\ &\quad + EP\{T_0' \neq 1 \mid X_{1n}(1,1), X_{2n}(1,1)\} \\ &= P_1 + P_2 + P_3 + P_4, \text{ say.} \end{aligned} \tag{14.2.1}$$

We now bound P_1. This follows from an immediate application of Theorem 7.3.1. We obtain

$$P_1 \leqq \frac{1}{\alpha_1} + P\{I(X_{1n}, Z_n \mid U_n, X_{2n}) \leqq \log \alpha_1 M_1\}\,. \tag{14.2.2}$$

In exactly the same way

$$P_2 \leqq \frac{1}{\alpha_2} + P\{I(X_{2n}, Z_n \mid U_n, X_{1n}) \leqq \log \alpha_2 M_2\}\,. \tag{14.2.3}$$

To bound P_3, note that $T_0' = 1$, and $(M_1 - 1)(M_2 - 1)$ is the number of ways to make the corresponding wrong decision. Then (X_1, X_2) plays the same role that X_1 and X_2 separately played in (14.2.2) and (14.2.3), and we obtain that

$$P_3 \leqq \frac{1}{\alpha_3} + P\{I([X_{1n}, X_{2n}], Z_n \mid U_n) \leqq \log \alpha_3 M_1 M_2\}\,. \tag{14.2.4}$$

Finally, to bound P_4, note that $(M_0 - 1) M_1 M_2$ is the number of ways to make the corresponding wrong decision, and that no variable is fixed because both $u_1(1, 1)$ and $u_2(1, 1)$ are wrongly decoded, even when the second indices are correctly decoded. Hence

$$P_4 \leqq \frac{1}{\alpha_4} + P\{I([X_{1n}, X_{2n}], Z_n) \leqq \log \alpha_4 M_0 M_1 M_2\}\,. \tag{14.2.5}$$

For simplicity set $\alpha_1 = \alpha_2 = \alpha_3 = \alpha_4 = \exp_2\{s\sqrt{n}\}$, $s > 0$, and $\log M_0$, $\log M_1$, $\log M_2$ any non-negative numbers such that

$$0 \leqq \log M_1 \leqq n\bar{I}\,(X_1, Z \mid U, X_2) - 2s\sqrt{n}\,. \tag{14.2.6}$$

$$0 \leqq \log M_2 \leqq n\bar{I}\,(X_2, Z \mid U, X_1) - 2s\sqrt{n}\,. \tag{14.2.7}$$

$$0 \leqq \log M_1 + \log M_2 \leqq n\bar{I}\,([X_1, X_2], Z \mid U) - 2s\sqrt{n}\,. \tag{14.2.8}$$

$$0 \leqq \log M_0 + \log M_1 + \log M_2 \leqq n\bar{I}\,([X_1, X_2], Z) - 2s\sqrt{n}\,. \tag{14.2.9}$$

Applying Chebyshev's inequality for s sufficiently large to (14.2.2)–(14.2.5), as in (7.3.14), we can make each of P_1, P_2, P_3, and P_4 as small as desired. This proves the theorem.

Actually, we have proved the theorem with $0(\sqrt{n})$, not merely $o(n)$, in the exponent of the code length. However, since we will now prove a weak, not a strong, converse, there is no point in emphasizing this slight improvement.

14.3. Converse of the coding theorem. We now prove

Theorem 14.3.1. Let $\delta > 0$ be arbitrary, and λ, depending only upon δ, be sufficiently small. For any n, the rate triple

$$\left(\frac{1}{n}\log M_0,\ \frac{1}{n}\log M_1,\ \frac{1}{n}\log M_2\right)$$

of a code with $\bar{\lambda} \leqq \lambda$ is at a distance less than δ from R^*.

Proof: This consists simply of repeated use of Theorem 7.4.1. First define, for the code under consideration,

$$P'_1 = P\{T'_1 \neq T_1\}$$

$$P'_2 = P\{T'_2 \neq T_2\}$$

$$P'_3 = P\{T'_1 \neq T_1 \text{ or } T'_2 \neq T_2\}$$

$$P'_4 = P\{T'_0 \neq T_0 \text{ or } T'_1 \neq T_1 \text{ or } T'_2 \neq T_2\}\,.$$

Then Theorem 7.4.1 gives immediately, writing

$$h(\alpha) = -\alpha\log\alpha - (1-\alpha)\log(1-\alpha)$$

for $0 \leqq \alpha \leqq 1$,

$$H\,(u_1(T_0, T_1) \mid Z_n, T_0, T_2) \leqq P'_1\log(M_1 - 1) + h(P'_1)\,. \tag{14.3.1}$$

$$H\,(u_2(T_0, T_2) \mid Z_n, T_0, T_1) \leqq P'_2\log(M_2 - 1) + h(P'_2)\,. \tag{14.3.2}$$

In the situation which corresponds to P'_3, the wrong decision can be made in $(M_1M_2 - 1)$ ways. Hence

$$H(u_1(T_0, T_1), u_2(T_0, T_2) \mid Z_n, T_0) \leqq P'_3\log(M_1M_2 - 1) + h(P'_3)\,. \tag{14.3.3}$$

Finally, in the situation which corresponds to P'_4, the wrong decision can be made in $(M_0 M_1 M_2 - 1)$ ways. Hence

$$H(u_1(T_0, T_1), u_2(T_0, T_2) | Z_n) \leqq P'_4 \log (M_0 M_1 M_2 - 1) + h(P'_4) . \quad (14.3.4)$$

Recalling that

$$H(u_1(T_0, T_1) | T_0, T_2) = \log M_1 , \quad (14.3.5)$$

$$H(u_2(T_0, T_2) | T_0, T_1) = \log M_2 , \quad (14.3.6)$$

$$H(u_1(T_0, T_1), u_2(T_0, T_2) | T_0) = \log M_1 M_2 , \quad (14.3.7)$$

$$H(u_1(T_0, T_1), u_2(T_0, T_2)) = \log M_0 M_1 M_2 , \quad (14.3.8)$$

we obtain, subtracting (14.3.1)–(14.3.4) from (14.3.5)–(14.3.8), that

$$(1 - P'_1) \log M_1 \leqq h(P'_1) + \bar{I}(u_1(T_0, T_1), Z_n | T_0, T_2) , \quad (14.3.9)$$

$$(1 - P'_2) \log M_2 \leqq h(P'_2) + \bar{I}(u_2(T_0, T_2), Z_n | T_0, T_1) , \quad (14.3.10)$$

$$(1 - P'_3) \log M_1 M_2 \leqq h(P'_3) + \bar{I}([u_1(T_0, T_1), u_2(T_0, T_2)], Z_n | T_0) , \quad (14.3.11)$$

$$(1 - P'_4) \log M_0 M_1 M_2 \leqq h(P'_4) + \bar{I}([u_1(T_0, T_1), u_2(T_0, T_2)], Z_n) . \quad (14.3.12)$$

Define

$$u_1(i, j) = (Y_{11}(i, j), \ldots, Y_{1n}(i, j)) \quad (14.3.13)$$

$$u_2(i, k) = (Y_{21}(i, k), \ldots, Y_{2n}(i, k)) .$$

Then

$$\begin{aligned}
&\bar{I}(u_1(T_0, T_1), Z_n | T_0, T_2) \\
&= \sum_{i=1}^{n} [H(Z^i | T_0, T_2, Z_{i-1}) \\
&\quad - H(Z^i | u_1(T_0, T_1), T_0, T_2, Z_{i-1})] \\
&= \sum_{i=1}^{n} [H(Z^i | T_0, T_2, Z_{i-1}) \qquad (14.3.14) \\
&\quad - H(Z^i | Y_{1i}(T_0, T_1), Y_{2i}(T_0, T_2))] \\
&\leqq \sum_{i=1}^{n} [H(Z^i | T_0, T_2) - H(Z^i | Y_{1i}(T_0, T_1), Y_{2i}(T_0, T_2))] \\
&= \sum_{i=1}^{n} \bar{I}(Z^i, Y_{1i}(T_0, T_1) | Y_{2i}(T_0, T_2)) .
\end{aligned}$$

Now define a Markov chain V_0, (V_1, V_2), V_3 as follows:

$$P\{V_0 = (k, i)\} = \frac{1}{nM_0}, \quad k = 1, \ldots, n; \; i = 1, \ldots, M_0. \tag{14.3.15}$$

When $V_0 = (k, i)$, V_1 and V_2 are independently distributed as follows:

$$P\{V_1 = Y_{1k}(i, j), V_2 = Y_{2k}(i, l)\} = \frac{1}{M_1 M_2},$$

$$j = 1, \ldots, M_1; \qquad l = 1, \ldots, M_2. \tag{14.3.16}$$

When $V_1 = Y_{1k}(i, j)$ and $V_2 = Y_{2k}(i, l)$, V_3 has the distribution

$$w(\cdot \mid Y_{1k}(i, j), Y_{2k}(i, l)). \tag{14.3.17}$$

Let us verify that V_0, (V_1, V_2), V_3 fulfill the conditions on U, (X_1, X_2), Z of Section 14.1. V_1 and V_2 are independent, given V_0. V_1 takes values in A, V_2 in B. When V_0, V_1, and V_2 are given, V_3 has the distribution $w(\cdot \mid V_1, V_2)$. From (14.3.14) we obtain that

$$\frac{1}{n} \bar{I}(u_1(T_0, T_1), Z_n \mid T_0, T_2) \leqq \bar{I}(V_1, V_3 \mid V_0, V_2). \tag{14.3.18}$$

Thus (14.3.9) and (14.3.18) give us a bound on $\frac{1}{n} \log M_1$. Similar arguments give us the corresponding bounds on $\frac{1}{n} \log M_2$, $\frac{1}{n} \log M_1 M_2$, and $\frac{1}{n} \log M_0 M_1 M_2$. Since

$$\bar{\lambda} = P'_4 \geqq P'_3 \geqq \max(P'_1, P'_2),$$

and $\bar{\lambda}$ can be chosen sufficiently small, the theorem follows.

14.4. Miscellaneous remarks. Fix U at some value u. To any point $w_1(\cdot \mid u)$, $w_2(\cdot \mid u)$ in $(a + b)$ – space make correspond the following point in $(ab + 3)$ – space: Its first ab coordinates are the values of $w_1(i \mid u) w_2(j \mid u)$, $i = 1, \ldots, a; j = 1, \ldots, b$. (These coordinates sum to one.) The next three coordinates are

$$\bar{I}(X_1, Z \mid X_2, U = u), \qquad \bar{I}(X_2, Z \mid X_1, U = u),$$

$$\bar{I}([X_1, X_2], Z \mid U = u).$$

By the standard argument based on Carathéodory's theorem we conclude that it suffices to take $d = ab + 2$. R^* can be computed to any desired degree of accuracy.

The results of this chapter are due to Slepian and Wolf [2]. Their proofs have been considerably streamlined. The problem where there is no sender 0 was studied by Ahlswede [2] and Liao [1].

15. Degraded Broadcast Channels

15.1. Formulation of the problem. Let $A = \{1, \ldots, a\}$, $B = \{1, \ldots, b\}$, and $C = \{1, \ldots, c\}$. Let $w_1(\cdot|\cdot)$ be the c.p.f. of a d.m.c. $A \to B$, and $w_2(\cdot|\cdot)$ the c.p.f. of a d.m.c. $B \to C$. Write A_n^* for the Cartesian product of n A's, with similar definitions for B_n^*, C_n^*, and, later, for D_n^*. There are three individuals I, II, III who act as follows: I wishes to send to II an arbitrarily chosen pair (i, j) of messages over the d.m.c. $w_1(\cdot|\cdot)$, and the message j to III over the d.m.c. $w_3(\cdot|\cdot)$, where

$$w_3(s_2|s_1) = \sum_{s=1}^{b} w_2(s_2|s)\, w_1(s|s_1)$$

$$1 \leqq s_1 \leqq a,\ 1 \leqq s_2 \leqq c\,.$$

I therefore sends a properly chosen n-sequence $x_{ij} \varepsilon A_n^*$, which is received by II as the chance n-sequence $v_1(x_{ij}) \varepsilon B_n^*$. This last sequence is automatically transmitted to III over the d.m.c. $w_2(\cdot|\cdot)$, and received by him as the chance n-sequence $v_2(v_1(x_{ij})) \varepsilon C_n^*$. The set $\{1, \ldots, M_1\}$ is the set of possible messages i, and the set $\{1, \ldots, M_2\}$ is the set of possible messages j. The disjoint decoding sets A_{ij} of II are to satisfy

$$P\{v_1(x_{ij}) \varepsilon A_{ij}\} \geqq 1 - \lambda_1, \quad 1 \leqq i \leqq M_1, \quad 1 \leqq j \leqq M_2\,.$$

The disjoint decoding sets B_j of III are to satisfy

$$P\{v_2(v_1(x_{ij})) \varepsilon B_j\} \geqq 1 - \lambda_2, \quad 1 \leqq j \leqq M_2\,.$$

Of course, $0 < \lambda_1 < 1$, $0 < \lambda_2 < 1$.

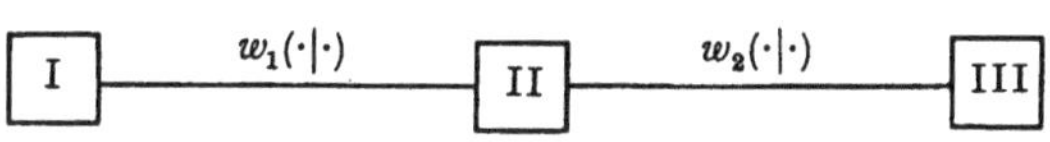

A point (r_1, r_2) is called (λ_1, λ_2)-achievable or (λ_1, λ_2)-attainable if, for $\varepsilon > 0$ arbitrary and n sufficiently large, there exists a code as above, i.e., the sequences x_{ij} and decoding sets A_{ij} and B_j, such that

$$M_1 > \exp_2\{n(r_1 - \varepsilon)\}, \quad M_2 > \exp_2\{n(r_2 - \varepsilon)\}\,. \qquad (15.1.1)$$

A point (r_1, r_2) is said to be achievable or attainable if it is achievable or attainable for every (λ_1, λ_2). The problem of this chapter is to obtain the regions $\bar{R}(\lambda_1, \lambda_2)$ and $\bar{R}$ of all (λ_1, λ_2)-achievable and achievable points, respectively. It will be shown that all the regions are the same.

In our coding theorem we can actually attain $0(\sqrt{n})$ in the exponent of the expression for the length. We have used the definition of (15.1.1) because the methods used do not permit an $0(\sqrt{n})$ term in the converse.

15.2. A coding theorem. Let $D = \{1, \ldots, d\}$, and $w(\cdot|\cdot)$ be a c.p.f. $D \to A$. Let the chance variables U, X, Y, Z form, in this order, a Markov chain, as follows: U has the distribution π on D. Also

$$P\{X = j | U = i\} = w(j|i), \ i = 1, \ldots, d; j = 1, \ldots, a$$

$$P\{Y = j | X = i\} = w_1(j|i), \ i = 1, \ldots, a; j = 1, \ldots, b$$

$$P\{Z = j | Y = i\} = w_2(j|i), \ i = 1, \ldots, b; j = 1, \ldots, c\,.$$

Let R^{**} be the totality of pairs

$$\bar{I}(X, Y | U)\,, \ \bar{I}(U, Z) \tag{15.2.1}$$

for all d, π, and w. Let R^* be the m-closure of R^{**}. We will now prove

Theorem 15.2.1. $\quad \bar{R} \supseteq R^*$.

The original proof of this theorem by Bergmans uses the method of random codes. It is of interest and utility to show that the method used by us in Chapters 3, 4, and elsewhere can be modified to prove this theorem, and this is how we will proceed.

Let (U^i, X^i, Y^i, Z^i), $i = 1, \ldots, n$ be independent chance variables with the same distribution as (U, X, Y, Z). Write $U_n = (U^1, \ldots, U^n)$, and similarly for X_n, Y_n, and Z_n. Let $\varepsilon > 0$ be a small number which we will choose later. By proper choices of multiplicative constants we can achieve the following:

$$P\{X_n \text{ is generated by } u_n | U_n = u_n\} > 1 - \frac{\varepsilon}{4},$$

$$P\{Y_n \text{ is generated by } x_n | X_n = x_n\} > 1 - \frac{\varepsilon}{4},$$

$$P\{Z_n, \text{ is generated by } u_n | U_n = u_n\} > 1 - \frac{\varepsilon}{4}.$$

Let $\{(u_1, B_1), \ldots, (u_{M_2}, B_{M_2})\}$ be a code (n, M_2, ε) such that all the u's are πn-sequences, the sets B_j contain only n-sequences generated $U \to Z$ by u_j, and

$$M_2 > \exp_2\{n\bar{I}(U, Z) - \sqrt{n}K(\varepsilon)\}\,. \tag{15.2.2}$$

Let $v_0(u_j)$ be the chance received sequence in A_n^* when $u_j \varepsilon D_n^*$ is sent over the d.m.c. w. Write $v(u_j) = v_2(v_1(v_0(u_j)))$. A path from u_j to any point z_n in B_j can be described thus: $u_j \to v_0(u_j) \to v_1(v_0(u_j)) \to v_2(v_1(v_0(u_j))) = z_n$. Define G_j as the set of n-sequences in A_n^* generated by u_j. Then,

$$P\{v(u_j) \varepsilon B_j, v_0(u_j) \varepsilon G_j\} > 1 - 2\varepsilon\,. \tag{15.2.3}$$

For each $x_n \varepsilon A_n^*$ define

$$\gamma_j(x_n) = P\{Z_n \varepsilon B_j | X_n = x_n\} . \tag{15.2.4}$$

Then, from (15.2.3),

$$\sum_{x_n \varepsilon G_j} \gamma_j(x_n) P\{v_0(u_j) = x_n\} > 1 - 2\varepsilon . \tag{15.2.5}$$

Choose ε so small that $4\sqrt{\varepsilon} \leqq \min(\lambda_1, \lambda_2)$, and set $t = \frac{2}{\sqrt{\varepsilon}} - 1$. Define

$$\mu_j = P\{v_0(u_j) \varepsilon G_j, \gamma_j(v_0(u_j)) \leqq 1 - 2(t+1)\varepsilon\} . \tag{15.2.6}$$

Then, from (15.2.5),

$$\mu_j [1 - 2(t+1)\varepsilon] + 1 - \mu_j > 1 - 2\varepsilon , \tag{15.2.7}$$

so that

$$\mu_j < \frac{1}{t+1} = \frac{\sqrt{\varepsilon}}{2} . \tag{15.2.8}$$

From G_j we now delete all points x_n for which $\gamma_j(x_n) \leqq 1 - 2(t+1)\varepsilon$; call the resulting set G_j'. For $x_n \varepsilon G_j'$, $\gamma_j(x_n) > 1 - 2(t+1)\varepsilon = 1 - 4\sqrt{\varepsilon}$, so that, when ε is sufficiently small, as will be the case, the various G_j' are disjoint. Using Theorem 7.2.1 and its corollary for the sets G_j', we conclude that, for each j, $1 \leqq j \leqq M_2$, there exists a code

$$\{(x_{1j}, A_{1j}), (x_{2j}, A_{2j}), \ldots, (x_{M_1 j}, A_{M_1 j})\}$$

such that the x_{ij} are elements of G_j', the disjoint $A_{ij} \subset B_n^*$ contain only sequences generated by x_{ij} and satisfy

$$P\{v_1(x_{ij}) \varepsilon A_{ij}\} > 1 - 3\sqrt{\varepsilon} , \tag{15.2.9}$$

and

$$P\{v_2(v_1(x_{ij})) \varepsilon B_j\} > 1 - 2(t+1)\varepsilon = 1 - 4\sqrt{\varepsilon} \tag{15.2.10}$$

for $1 \leqq i \leqq M_1$, where

$$M_1 > \exp_2\{n\bar{I}(X, Y | U) - \sqrt{n} K(\varepsilon)\} . \tag{15.2.11}$$

The theorem would now be proved if it were known that the A_{ij} are disjoint. For $i \neq i'$, A_{ij} and $A_{i'j}$ are always disjoint, by the construction of the codes. It remains to consider A_{ij} and $A_{i'j'}$ when $j \neq j'$. For this purpose define, for every $y_n \varepsilon B_n^*$,

$$\gamma_j'(y_n) = P\{Z_n \varepsilon B_j | Y_n = y_n\} . \tag{15.2.12}$$

Then, from (15.2.9) and (15.2.10),

$$\sum_{y_n \varepsilon A_{ij}} \gamma_j'(y_n)\, P\{v_1(x_{ij}) = y_n\} > 1 - 7\sqrt{\varepsilon}\,. \qquad (15.2.13)$$

Define

$$\mu_{ij}' = P\{v_1(x_{ij})\, \varepsilon A_{ij},\, \gamma_j'(v_1(x_{ij})) \leqq 1 - \sqrt[4]{\varepsilon}\}\,. \qquad (15.2.14)$$

Then, from (15.2.13),

$$\mu_{ij}'[1 - \sqrt[4]{\varepsilon}] + 1 - \mu_{ij}' > 1 - 7\sqrt{\varepsilon}\,. \qquad (15.2.15)$$

so that

$$\mu_{ij}' < 7\sqrt[4]{\varepsilon}\,. \qquad (15.2.16)$$

From A_{ij} now delete all points y_n for which $\gamma_j'(y_n) \leqq 1 - \sqrt[4]{\varepsilon}$; call the resulting set A_{ij}'. Then

$$P\{v_1(x_{ij})\, \varepsilon A_{ij}'\} > 1 - 3\sqrt{\varepsilon} - 7\sqrt[4]{\varepsilon} \qquad (15.2.17)$$

by (15.2.16) and (15.2.9), while (15.2.10) remains unchanged. We now decrease ε even further, if necessary, so that

$$0 < 10\sqrt[4]{\varepsilon} < \min\left(\frac{1}{2}, \lambda_1, \lambda_2\right). \qquad (15.2.18)$$

Suppose y_n belonged to $A_{i'j'}'$ and A_{ij}', $j \neq j'$. Then $\gamma_j'(y_n) > 1 - \sqrt[4]{\varepsilon} > \frac{1}{2}$, $\gamma_{j'}'(y_n) > \frac{1}{2}$. This is impossible, since B_j and $B_{j'}$ are disjoint, by the code construction. The desired code has been constructed, thus:

$$\{x_{ij}, A_{ij}', B_j\},\ i = 1, \ldots, M_1;\ j = 1, \ldots, M_2, \qquad (15.2.19)$$

and the theorem is proved.

Actually it is not even necessary to have the disjoint sets A_{ij}'. Simply write

$$A_i = \bigcup_j A_{ij}'. \qquad (15.2.20)$$

Then the A_i are disjoint, and the code

$$\{x_{ij}, A_i, B_j\},\ i = 1, \ldots, M_1;\ j = 1, \ldots, M_2 \qquad (15.2.21)$$

would be a desired code, except for one thing: How is individual II to decode the message j? The answer is this: He performs a chance experiment with c.p.f. $w_2(\cdot\,|\,\cdot)$ on the sequence he receives. After all, this, is exactly what the channel w_2 does. His decision on the message j is made on the basis of which B_j the chance "received" sequence lies in.

15.3. Beginning of the proof of the strong converse. This section is devoted to the proof of

Theorem 15.3.1. The following is true:

$$\overline{R} = \overline{R}(\lambda_1, \lambda_2)\,. \tag{15.3.1}$$

In the next section we will prove the weak converse

$$R^* \geqq \overline{R}\,. \tag{15.3.2}$$

The two statements, together with Theorem 15.2.1, imply the coding theorem and strong converse,

$$\overline{R} = \overline{R}(\lambda_1, \lambda_2) = R^*\,.$$

Proof: Suppose given a code with the notation of (15.2.21). Let $\{k_n\}$ be a sequence of positive integers such that $k_n \cdot n^{-1} \to 0$ and $k_n \cdot n^{-1/2} \to \infty$. Let Γ be the operation defined in Section 7.9, and write

$$\Gamma^{k_n} A_j = A_i',\ \Gamma^{k_n} B_j = B_j'\,.$$

From Theorem 7.9.2 it follows that

$$P\{v_1(x_{ij})\,\varepsilon\,A_i'\} > 1 - \varepsilon_n \tag{15.3.3}$$

and

$$P\{v_2(v_1(x_{ij}))\,\varepsilon\,B_j'\} > 1 - \varepsilon_n\,, \tag{15.3.4}$$

where $\varepsilon_n \to 0$ as $n \to \infty$, uniformly for all (λ_1, λ_2)-codes, $\lambda_1 > 0, \lambda_2 > 0$, both fixed. Define

$$S_1(y_n) = \{i \,|\, y_n\,\varepsilon\,A_i'\}\,, \tag{15.3.5}$$

$$S_2(z_n) = \{j \,|\, z_n\,\varepsilon\,B_j'\}\,. \tag{15.3.6}$$

Now

$$y_n\,\varepsilon\,A_i' \Leftrightarrow A_i \cap \Gamma^{k_n}\{y_n\} \neq \phi\,. \tag{15.3.7}$$

Since the A_i's are disjoint we have

$$\#\,(S_1(y_n)) \leqq \#\,(\Gamma^{k_n}\{y_n\})\,. \tag{15.3.8}$$

Also, recalling the definition of b in Section 15.1, we have

$$\#\,(\Gamma^{k_n}\{y_n\}) \leqq \sum_{i=0}^{k_n} \binom{n}{i} b^i \leqq b^{k_n}(k_n+1)\binom{n}{k_n} \leqq 2^{n\delta_n} \tag{15.3.9}$$

where the second inequality is valid when $k_n < \frac{n}{2}$ (which is true for large n), and $\delta_n \to 0$ as $n \to \infty$, by Stirling's formula and the fact that $x \log x \to 0$ as $x \to 0$. The same argument gives that

$$\#\,(S_2(z_n)) \leqq 2^{n\delta_n}. \tag{15.3.10}$$

Let U be a chance variable, uniformly distributed on the set of values $\{1, \ldots, M_2\}$. Define

$$T(j) = \{x_{ij}, 1 \leqq i \leqq M_1\}.$$

Let X_n be a chance variable distributed as follows: When $U = j$, X_n is uniformly distributed on $T(j)$. Then

$$\log M_2 = H(U) = \bar{I}(U, v_2(v_1(X_n))) + H(U \mid v_2(v_1(X_n))). \tag{15.3.11}$$

Define, for $z_n \varepsilon C_n^*$,

$$\mu_n = E\,P\{U \notin S_2(v_2(v_1(X_n))) \mid v_2(v_1(X_n))\}.$$

Then $\mu_n \leqq \varepsilon_n$ and hence $\to 0$ as $n \to \infty$. From (7.4.1) and (15.3.10) we have

$$H\,(U \mid v_2(v_1(X_n))) \leqq 1 + \mu_n \log M_2 + (1 - \mu_n)\,n\delta_n. \tag{15.3.12}$$

Substituting this into (15.3.11) we obtain

$$\begin{aligned} \frac{1}{n} \log M_2 &\leqq \frac{1}{n}\,[\bar{I}(U, v_2(v_1(X_n))) + 1 + \mu_n \log M_2 + (1 - \mu_n)\,n\delta_n] \\ &= \frac{1}{n}\,[\bar{I}(U, v_2(v_1(X_n)))] + o(1). \end{aligned} \tag{15.3.13}$$

Also

$$\begin{aligned} \bar{I}(X_n, v_1(X_n) \mid U) &= H(X_n \mid U) - H\,(X_n \mid U, v_1(X_n)) \\ &= \log M_1 - H\,(X_n \mid U, v_1(X_n)). \end{aligned} \tag{15.3.14}$$

From (7.4.1), (15.3.8) and (15.3.9) we obtain

$$H(X_n \mid U, v_1(X_n)) \leqq 1 + \mu_n \log M_1 + (1 - \mu_n)\,n\delta_n. \tag{15.3.15}$$

From (15.3.14) and (15.3.15) we obtain

$$\begin{aligned} \frac{1}{n} \log M_1 \leqq \frac{1}{n}\,\bar{I}(X_n, v_1(X_n) \mid U) + \frac{1}{n} \\ + \frac{\mu_n}{n} \log M_1 + \frac{(1 - \mu_n)}{n} \cdot n\delta_n. \end{aligned} \tag{15.3.16}$$

Hence

$$\frac{1}{n}\log M_1 \leqq \frac{1}{n}\bar{I}(X_n, v_1(X_n)\,|\,U) + o(1)\,. \tag{15.3.17}$$

The $o(1)$ terms in (15.3.13) and (15.3.17) approach zero as $n \to \infty$, uniformly in all (λ_1, λ_2)-codes, $\lambda_1 > 0$, $\lambda_2 > 0$, both fixed. We now prove

Lemma 15.3.1. The pair

$$\frac{1}{n}\bar{I}\left(X_n, v_1(X_n)\,|\,U\right), \ \frac{1}{n}\bar{I}\left(U, v_2(v_1(X_n))\right) \tag{15.3.18}$$

belongs to $\overline{R}$.

Proof: We can use the pair (15.3.18) exactly as the pair (15.2.1) was used in the proof of Theorem 15.2.1. The new letters (except the u's) are now blocks of n letters, whose length is taken into account by the factor $\frac{1}{n}$. We conclude that, for sufficiently large multiples mn of n, there exist codes with rate pair within ε of (15.3.18) for any positive pair (λ_1, λ_2). For sufficiently large integers of the form $mn + n'$, $0 < n' < n$, one simply "wastes" the last, say, n' letters by not using them for the message. This proves the lemma.

It is tempting to assert that we have actually proved that the pair (15.3.18) belongs to R^*. However, the "dimensions" of the letters are not right for R^*. We will make them right by introducing U^i in the next section. Then we will prove that (15.3.18) belongs to R^*.

Suppose now that there exists a point $(\alpha_1, \alpha_2)\,\varepsilon\,\overline{R}(\lambda_1, \lambda_2)$ which is not in $\overline{R}$. For an increasing sequence of n's we construct a sequence of codes such that the rate point

$$\frac{1}{n}\log M_1\,, \ \frac{1}{n}\log M_2$$

converges to (α_1, α_2). From (15.3.13), (15.3.17), Lemma 15.3.1, and the compactness of $\overline{R}$, we conclude that there is a point (β_1, β_2) in $\overline{R}$ such that

$$\alpha_1 \leqq \beta_1, \quad \alpha_2 \leqq \beta_2\,.$$

This is impossible. Since $\overline{R}$ is m-closed, (α_1, α_2) would be in $\overline{R}$, contrary to the assumption. Thus we have proved the theorem.

15.4. Proof of the weak converse. This section is devoted to the proof of

Theorem 15.4.1. The following inequality holds for all λ_1, λ_2, both positive:

$$R^* \supseteqq \overline{R}(\lambda_1, \lambda_2)\,. \tag{15.4.1}$$

Proof: Actually it would be enough to prove the seemingly weaker result

$$R^* \supseteqq \bar{R}\,, \tag{15.4.2}$$

because of Theorem 15.3.1. However, (15.4.1) is easy enough to prove.

In the notation of the last section after (15.3.10), write out $v_1(X_n)$ as

$$v_1(X_n) = (Y^1, \ldots, Y^n) = Y_n, \tag{15.4.3}$$

and define

$$U^i = (U, Y^1, \ldots, Y^{i-1}), \quad i = 1, \ldots, n\,. \tag{15.4.4}$$

Also write out $v_2\,(v_1(X_n))$ as

$$v_2\,(v_1(X_n)) = (Z^1, \ldots, Z^n) = Z_n. \tag{15.4.5}$$

Let X^i denote the ith component of X_n. We note that, for every i, U^i, X^i, Y^i, Z^i form, in this order, a Markov chain.

Temporarily assume the truth of the following lemmas:

Lemma 15.4.1. We have

$$\bar{I}(U, Z_n) \leqq \sum_{i=1}^{n} \bar{I}(U^i, Z^i)\,. \tag{15.4.6}$$

Lemma 15.4.2. We have

$$\bar{I}(X_n, Y_n \mid U) = \sum_{i=1}^{n} \bar{I}(X^i, Y^i \mid U^i)\,. \tag{15.4.7}$$

From (15.3.13), (15.3.17), the above lemmas, and the fact that R^* is convex and m-closed, Theorem 15.4.1 follows at once.

Proof of Lemma 15.4.1: We have

$$\begin{aligned} H(Z_n \mid U) &= \sum_{i=1}^{n} H(Z^i \mid U, Z^1, \ldots, Z^{i-1}) \\ &\geq \sum_{i=1}^{n} H(Z^i \mid U, Y^1, \ldots, Y^{i-1}) \\ &= \sum_{i=1}^{n} H(Z^i \mid U^i)\,. \end{aligned} \tag{15.4.8}$$

Hence

$$\bar{I}(U, Z_n) \leqq \sum_{i=1}^{n} H(Z^i) - H(Z_n | U) \tag{15.4.9}$$

$$\leqq \sum_{i=1}^{n} \bar{I}(U^i, Z^i),$$

where the last inequality follows from (15.4.8).

Proof of Lemma 15.4.2. We have

$$\bar{I}(X_n, Y_n | U) = H(Y_n | U) - H(Y_n | X_n, U), \tag{15.4.10}$$

$$H(Y_n | U) = \sum_{i=1}^{n} H(Y^i | U^i), \tag{15.4.11}$$

$$H(Y_n | X_n, U) = H(Y_n | X_n), \tag{15.4.12}$$

(since U, X_n, Y_n form a Markov chain)

$$= \sum_{i=1}^{n} H(Y^i | X^i) = \sum_{i=1}^{n} H(Y^i | X^i, U^i)$$

(since U^i, X^i, Y^i form a Markov chain). The lemma now follows from (15.4.10)–(15.4.12).

15.5. Miscellaneous remarks. The idea of degraded broadcast channels is due to Cover [*2*]. The coding Theorem 15.2.1 was proved by Bergmans [*1*], using the random coding theorem. Our proof is new.

The weak converse

$$\bar{R} \subseteq R^*$$

was conjectured by Bergmans [*1*]. Almost all of it was proved by Gallager [*2*], and subsequently Ahlswede and Körner [*1*] proved the rest. A very special case was proved earlier by Wyner and Ziv [*2*].

The strong converse (Theorem 15.3.1) was proved by Ahlswede, Gacs, and Körner [*1*, *2*]. Our proof differs from theirs beginning with our Lemma 15.3.1. Our proof of the weak converse is new.

R^* is a closed convex set which can be described as follows: Its boundary consists of the line from the origin to the point

$$(\bar{I}(X, Y), 0), \tag{15.5.1}$$

(which is the point (15.2.1) when U is independent of X), the line from the origin to the point

$$(0, \bar{I}(X, Z)), \tag{15.5.2}$$

(which is the point (15.2.1) with $U = X$), and the curve $t(x)$, $0 \leqq x \leqq \bar{I}(X, Z)$, defined as follows:

$$t(x) = \max \bar{I}(X, Y \mid U), \tag{15.5.3}$$

where the maximum is over all chance variables U such that U, X, Y, Z form, in this order, a Markov chain, and $\bar{I}(U, Z) \geqq x$.

For any value i of U let m be the point with the following coordinates: Its first a coordinates are

$$w(1 \mid i), \ldots, w(a \mid i).$$

Its next two and last coordinates are

$$\bar{I}(X, Y \mid U = i), H(Z \mid U = i).$$

By the standard argument based on Carathéodory's theorem we conclude that R^* is already obtained when $d = (a + 1)$, and that it can be computed to any desired degree of accuracy.

References

AHLSWEDE, R., and J. WOLFOWITZ: [1] Correlated decoding for channels with arbitrarily varying channel probability functions. Inform. and Control **14**, No. 5, 457—473 (1969). — [2] The capacity of a channel with arbitrarily varying channel probability functions and binary output alphabet. Zeitschr. f. Wahrsch. u.v.G. **15**, No. 3, 186—194 (1970).

AHLSWEDE, R.: [1] Lecture at the Math. Forschungsinstitut in Oberwolfach, Germany, on March 21, 1977. — [2] Multiway communication channels. Proc. 2nd Internat. Symp. on Info. Transmission, Tsakhkasdor, Armenia, U.S.S.R., Hungarian Press, 1971.

AHLSWEDE, R., P. GÁCS and J. KÖRNER: [1] Bounds on conditional probabilities with applications in multi-user communication. Zeitschr. f. Wahrsch. u.v.G. **34**, 157—177 (1976). — [2] Correction to [1], ibid., **39**, 353—4 (1977).

AHLSWEDE, R., and G. DUECK: [1] Every bad code has a good subcode: A local converse to the coding theorem. Zeitsch. f. Wahrsch. u.v.G. **34**, 179—182 (1976).

AHLSWEDE, R., and J. KÖRNER: [1] Source coding with side information and a converse for degraded broadcast channels. IEEE Trans. Info. Theory, IT-**21**, 629—637 (1975).

BERGMANS, P. P.: [1] Random coding theorem for broadcast channels with degraded components. IEEE Trans. Info. Theory, IT-**19**, 197—207 (1973).

BLACKWELL, D., L. BREIMAN and A. J. THOMASIAN: [1] Proof of Shannon's transmission theorem for finite-state indecomposable channels. Ann. Math. Stat. **29**, No. 4, 1209—1220 (1958). — [2] The capacity of a class of channels. Ann. Math. Stat. **30**, No. 4, 1229—1241 (1959).

BREIMAN, L.: [1] The individual ergodic theorem of information theory. Ann. Math. Stat. **28**, No. 3, 809—811 (1957). — [2] Correction to [1]. Ann. Math. Stat. **31**, No. 3, 809—810 (1960).

COVER, T. M.: [1] A proof of the data compression theorem of Slepian and Wolf for ergodic sources. IEEE Trans. Info. Theory, IT-**21**, No. 2, 226—228 (1975). — [2] Broadcast channels. IEEE Trans. Info. Theory, IT-**18**, 2—14 (1972).

DOBRUSHIN, R. L., and S. Z. STAMBLER: [1] Coding theorems for arbitrarily varying discrete memoryless channels. Problemi Peredachi Informatsii, **11**, No. 2, 3—22 (1975). Russian.

EGGLESTON, H. G.: Convexity. Cambridge University Press, 1958.

ELIAS, P.: [1] Coding for noisy channels. IRE Convention Record **4**, 37—44 (1955). — [2] Coding for two noisy channels. Proc. Third London Symposium on Information Theory. London: Butterworth Scientific Publications 1955.

FANO, R. M.: [1] Statistical theory of communication. Notes of a course given at M.I.T. 1952, 1954.

FEINSTEIN, A.: [1] A new basic theorem of information theory. Trans. IRE, PGIT, Sept. 1954, 2—22. — [2] Foundations of information theory. New York: McGraw-Hill Book Co. 1958. — [3] On the coding theorem and its converse for finite-memory channels. Inform. and Control **2**, No. 1, 25—44 (1959).

FELLER, W.: [1] An introduction to probability theory and its applications. Vol. 1, third edition. New York: John Wiley and Sons, 1968.

GALLAGER, R. G.: [1] Information theory and reliable communication. New York, John Wiley and Sons, 1968. — [2] Coding for degraded broadcast channels. Problemi Peredachi Informatsii. **10**, Nr. 3, 3—14 (1974). Russian.

GRAY, R. M., and A. D. WYNER: [1] Source coding for a simple network. Bell System Tech. Journal, **53**, No. 9, 1681—1721 (1974).

JACOBS, K.: [1] Almost periodic channels. Colloquium on combinatorial methods in probability theory, August 1—10, 1962, pp. 118—126. Matematisk Institute, Aarhus University.

KAKUTANI, S.: [1] A generalization of Brouwer's fixed point theorem. Duke Math. Journal **8**, 457—458 (1941).

KESTEN, H.: [1] Some remarks on the capacity of compound channels in the semi-continuous case. Inform. and Control **4**, 169—184 (1961).

KIEFER, J., and J. WOLFOWITZ: [1] Channels with arbitrary varying channel probability functions. Inform. and Control **5**, 44—54 (1962).

KHINCHIN, A. YA.: [1] On the fundamental theorems of information theory. Uspekhi Mat. Nauk **11**, No. 1 (67), 17—75 (1956).

LIAO, H.: [1] Multiple access channels. Ph. D. dissertation. Dep't of Electrical Engineering, University of Hawaii, Honolulu, 1972.

MARGULIS, G. A.: [1] Probabilistic characteristics of graphs with large connectivity. Problemi Peredachi Informatsii **10**, 101—108 (1974).Russian.

MCMILLAN, B.: [1] The basic theorems of information theory. Ann. Math. Stat. **24**, No. 2, 196—219 (1953).

SHANNON, C. E.: [1] A mathematical theory of communication. Bell System Tech. J. **27**, 379—423, 623—656 (1948). — [2] Certain results in coding theory for noisy channels. Inform. and Control **1**, 6—25 (1957). — [3] Channels with side information at the transmitter. IBM J. Research and Development **2**, No. 4, 289—293 (1958). — [4] Probability of error for optimal codes in a Gaussian channel. Bell System Tech. J. **38**, No. 3, 611—656 (1959). — [5] Coding theorems for a discrete source with a fidelity criterion. Information and decision processes, R. E. MACHOL, editor, MCGRAW-HILL, 1960. New York. — [6] On the zero-error capacity of a noisy channel. Trans. IRE, PGIT (1956), 8—19.

SLEPIAN, D., and J. K. WOLF: [1] Noiseless coding of correlated information sources. IEEE Trans. Info. Theory, IT-**19**, 471—480 (1973). — [2] A coding theorem for multiple access channels with correlated sources. Bell System Techn. Journal. **52**, No. 7, 1037—1076 (1973).

STAMBLER, S. Z.: [1] Shannon's theorems for a complete class of discrete channels, whose state is known to the receiver. Problemi Peredachi Informatsii **11**, 3—12 (1975). Russian.

STIGLITZ, I. G.: [1] Coding for a class of unkown channels. IEEE Trans. Info. Theory, IT-**12**, 189—195 (1966).

THOMASIAN, A. J.: [1] An elementary proof of the AEP of information theory. Ann. Math. Stat. **31**, No. 2, 452—456 (1960).

WEISS, L.: [1] On the strong converse of the coding theorem for symmetric channels without memory. Quart. Appl. Math. **18**, No. 3, 209—214 (1960).

WIENER, N.: [1] Cybernetics. New York: John Wiley and Sons **1948**.

WOLFOWITZ, J.: [1] The coding of messages subject to chance errors. Illinois J. Math. **1**, No. 4, 591—606 (1957). — [2] The maximum achievable length of an error correcting code. Illinois J. Math. **2**, No. 3, 454—458 (1958). — [3] Strong converse of the coding theorem for semi-continuous channels. Illinois J. Math. **3**, No. 4, 477—489 (1959). — [4] Simultaneous channels. Arch. Rational Mech. Anal. **4**, No. 4, 371—386 (1960). — [5] Strong converse of the coding theorem for the general discrete finite-memory channel. Inform. and Control **3**, No. 1, 89—93 (1960). — [6] On coding theorems for general simultaneous channels. Trans. IRE, PGCT, CT-**7**, No. 4, 513—516 (1960). — [7] On channels without capacity. Inform. and Control **6**, No. 1, 49—54 (1963). — [8] Approximation with a fidelity criterion. Proc. Fifth Berkeley Symp., 1966, 565—573. Berkeley and Los Angeles, University of California Press.

WYNER, A. D.: [1] On source coding with side information at the decoder. IEEE Trans. Info. Theory, IT-**21**, 294—300 (1975).

WYNER, A. D., and J. ZIV: [1] The rate distortion function for source coding with side information at the decoder. IEEE Trans. Info. Theory, IT-**22**, 1—10 (1976). — [2] A theorem on the entropy of certain binary sequences and applications. Part I. IEEE Trans. Info. Theory, IT-**19**, 769—772 (1973). Part II (by WYNER), ibid., 772—777.

Ergebnisse der Mathematik und ihrer Grenzgebiete

A Series of Modern Surveys in Mathematics

---

1. Bachmann: Transfinite Zahlen
2. Miranda: Partial Differential Equations of Elliptic Type
4. Samuel: Méthodes d'algèbre abstraite en géométrie algébrique
5. Dieudonné: La géométrie des groupes classiques
6. Roth: Algebraic Threefolds with Special Regard to Problems of Rationality
7. Ostmann: Additive Zahlentheorie. 1. Teil: Allgemeine Untersuchungen
8. Wittich: Neuere Untersuchungen über eindeutige analytische Funktionen
10. Suzuki: Structure of a Group and the Structure of its Lattice of Subgroups
11. Ostmann: Additive Zahlentheorie. 2. Teil: Spezielle Zahlenmengen
13. Segre: Some Properties of Differentiable Varieties and Transformations
14. Coxeter/Moser: Generators and Relations for Discrete Groups
15. Zeller/Beekmann: Theorie der Limitierungsverfahren
16. Cesari: Asymptotic Behavior and Stability Problems in Ordinary Differential Equations
17. Severi: Il teorema di Riemann-Roch per curve, superficie e varietà questioni collegate
18. Jenkins: Univalent Functions and Conformal Mapping
19. Boas/Buck: Polynomial Expansions of Analytic Functions
20. Bruck: A Survey of Binary Systems
21. Day: Normed Linear Spaces
23. Bergmann: Integral Operators in the Theory of Linear Partial Differential Equations
25. Sikorski: Boolean Algebras
26. Künzi: Quasikonforme Abbildungen
27. Schatten: Norm Ideals of Completely Continuous Operators
28. Noshiro: Cluster Sets
30. Beckenbach/Bellman: Inequalities
31. Wolfowitz: Coding Theorems of Information Theory
32. Constantinescu/Cornea: Ideale Ränder Riemannscher Flächen
33. Conner/Floyd: Differentiable Periodic Maps
34. Mumford: Geometric Invariant Theory
35. Gabriel/Zisman: Calculus of Fractions and Homotopy Theory
36. Putnam: Commutation Properties of Hilbert Space Operators and Related Topics
37. Neumann: Varieties of Groups
38. Boas: Integrability Theorems for Trigonometric Transforms
39. Sz.-Nagy: Spektraldarstellung linearer Transformationen des Hilbertschen Raumes
40. Seligman: Modular Lie Algebras
41. Deuring: Algebren
42. Schütte: Vollständige Systeme modaler und intuitionistischer Logik
43. Smullyan: First-Order Logic
44. Dembowski: Finite Geometries
45. Linnik: Ergodic Properties of Algebraic Fields
46. Krull: Idealtheorie
47. Nachbin: Topology on Spaces of Holomorphic Mappings
48. A. Ionescu Tulcea/C. Ionescu Tulcea: Topics in the Theory of Lifting
49. Hayes/Pauc: Derivation and Martingales
50. Kahane: Séries de Fourier absolument convergentes
51. Behnke/Thullen: Theorie der Funktionen mehrerer komplexer Veränderlichen
52. Wilf: Finite Sections of Some Classical Inequalities
53. Ramis: Sous-ensembles analytiques d'une variété banachique complexe
54. Busemann: Recent Synthetic Differential Geometry
55. Walter: Differential and Integral Inequalities
56. Monna: Analyse non-archimédienne
57. Alfsen: Compact Convex Sets and Boundary Integrals

58. Greco/Salmon: Topics in m-adic Topologies
59. López de Medrano: Involutions on Manifolds
60. Sakai: C^*-Algebras and W^*-Algebras
61. Zariski: Algebraic Surfaces
62. Robinson: Finiteness Conditions and Generalized Soluble Groups, Part 1
63. Robinson: Finiteness Conditions and Generalized Soluble Groups, Part 2
64. Hakim: Topos annelés et schémas relatifs
65. Browder: Surgery on Simply-Connected Manifolds
66. Pietsch: Nuclear Locally Convex Spaces
67. Dellacherie: Capacités et processus stochastiques
68. Raghunathan: Discrete Subgroups of Lie Groups
69. Rourke/Sanderson: Introduction of Piecewise-Linear Topology
70. Kobayashi: Transformation Groups in Differential Geometry
71. Tougeron: Idéaux de fonctions différentiables
72. Gihman/Skorohod: Stochastic Differential Equations
73. Milnor/Husemoller: Symmetric Bilinear Forms
74. Fossum: The Divisor Class Group of a Krull Domain
75. Springer: Jordan Algebras and Algebraic Groups
76. Wehrfritz: Infinite Linear Groups
77. Radjavi/Rosenthal: Invariant Subspaces
78. Bognár: Indefinite Inner Product Spaces
79. Skorohod: Integration in Hilbert Space
80. Bonsall/Duncan: Complete Normed Algebras
81. Crossley/Nerode: Combinatorial Functors
82. Petrov: Sums of Independent Random Variables
83. Walker: The Stone-Čech Compactification
84. Wells/Williams: Embeddings and Extensions in Analysis
85. Hsiang: Cohomology Theory of Topological Transformation Groups
86. Olevskii: Fourier Series with Respect to General Orthogonal Systems
87. Berg/Forst: Potential Theory on Locally Compact Abelian Groups
88. Weil: Elliptic Functions according to Eisenstein and Kronecker
89. Lyndon/Schupp: Combinatorial Group Theory
90. Edwards/Gaudry: Littlewood-Paley and Multiplier Theory
91. Gunning: Riemann Surfaces and Generalized Theta Functions
92. Lindenstrauss/Tzafriri: Classical Banach Spaces I
93. Besse: Manifolds all of whose Geodesics are Closed
94. Heyer: Probability Measures on Locally Compact Groups
95. Adian: The Burnside Problem and Identities in Groups

MIX
Papier aus verantwortungsvollen Quellen
Paper from responsible sources
FSC® C105338

If you have any concerns about our products,
you can contact us on
ProductSafety@springernature.com

In case Publisher is established outside the EU,
the EU authorized representative is:
Springer Nature Customer Service Center GmbH
Europaplatz 3, 69115 Heidelberg, Germany

Printed by Libri Plureos GmbH
in Hamburg, Germany